GROUNDWATER RECHARGE ESTIMATION AND WATER RESOURCES ASSESSMENT IN A TROPICAL CRYSTALLINE BASEMENT AQUIFER

Groundwater Recharge Estimation and Water Resources Assessment in a Tropical Crystalline Basement Aquifer

DISSERTATION

Submitted in fulfilment of the requirements of
the Board for Doctorates of Delft University of Technology
and of the Academic Board of the UNESCO-IHE Institute for Water Education for the
Degree of DOCTOR
to be defended in public
on Thursday, June 29, 2006 at 10:00 hours
in Delft, the Netherlands

by

Nyasha Lawrence NYAGWAMBO
born in Mutasa, Zimbabwe
Master of Science, UNESCO-IHE

This dissertation has been approved by the promotor
Prof. dr. ir. H.H.G. Savenije   TU Delft/UNESCO-IHE Delft, The Netherlands

Members of the Awarding Committee:
Chairman   Rector Magnificus TU Delft, The Netherlands
Co-chairman   Director UNESCO-IHE, The Netherlands
Prof. dr. S. Uhlenbrook   UNESCO-IHE, The Netherlands
Prof.dr. J. J. de Vries   Free University, The Netherlands
Prof.dr.ir. N. van de Giesen   TU Delft, The Netherlands
Prof.ir. T. Olsthoorn   TU Delft, The Netherlands
Dr. J. Sakupwanya   Zambesi Water Authority

Published by   A.A. Balkema Publishers, a member of Taylor & Francis Group plc
               www.balkema.nl and www.tandf.co.uk

ISBN 0 415 41692 2 / 978-0-415-41692-4 (Taylor & Francis Group)

To my family, Anna, Panashe, Ariko and The Three. To my Dad who says all the right things, my Mum who does most of the correct ones, and my sisters who cheer me all the way.

ACKNOWLEDGEMENTS

First and foremost I would like to thank my Promoter, Prof. Dr. Ir. Huub Savenije who kept on urging me on when the chips were down and invested so much of his time when my efforts seemed so hopeless. My special thanks too to Ir. Jan Nonner who sacrificed a lot of his time both in the Netherlands and in Zimbabwe to get this work were it is today. Then there was one very special person Marieke de Groen who kept on reminding me that the work could be done. Your encouragement was not in vain.

I would like to thank my colleagues at the University of Zimbabwe Civil Engineering Department for the encouragement and support. Evans Kaseke who always showed there was light at the end of the tunnel (even when it looked like a train!), David Katale whose discussions in other subjects gave me time off, Dr Bekithemba Gumbo who always reminded me of the "big picture", Dr Pieter van der Zaag, Dr Nhapi, Alexander Mhizha and of cause Dr Vassileva whose faith in me never wavered. Of cause the fieldwork would never have been carried out without Mssrs Chawira, Mbidzo and Mabika and the Chikara community's help. Thanks too to Maoneyi for calling me Dr even before the first draft was written, at least you kept the end in sight. To Thomas Natsa, Dr Patrick Moriarty and Dr Richard Owen those initial discussions were very insightful, thank you so much.

Most importantly, I would like to thank Anna and Panashe for letting me go away for all those months, my sisters for encouraging me all the way, and my parents for praying all the time. And to my superiors at IWSD, Eng Mudege and Mrs Neseni thanks for the encouragement and support when the load looked too heavy.

PREFACE

This study started in 1999. When I was asked to apply for a PhD study under the auspices of the Capacity Building Program for Zimbabwe and the Southern Africa Region I had no clue about groundwater let alone groundwater recharge as a science. This fact latter created problems for me during the course of the study. I was an engineer who branched to water resources management in my MSc study. So to the engineers the subject of groundwater recharge was like funky music to a monk – they may like it but not openly. I was embarking on a study that was neither hydrology no management, neither engineering nor science. In short I was disowned by all but then I had to do a PhD! I suppose this is what they call a Catch 22 situation.

But off I went. I wrote a proposal which I still suspect very few people read. Some commented with scorn and the hydrogeologists were so scared I wanted to work with them that they bolted at my sight. I felt lost and for a moment wanted to quit. But my Promoter kept probing. He too did not particularly like my topic but as usual he saw the future and urged me on. So I decided to do some work on recharge not because of its scientific appeal but because I believed, I am not so sure now, that it touches so many people in the third world, my village relatives included.

I always got fascinated by groundwater experts when they talk about sophisticated methods that are needed to tell people how much groundwater is available. Equally perplexing to me was the fact that so much is invested in determining accurately something so uncertain and imprecise as natural recharge. Surely the uncertainty in the final value did not warrant the investment. There must be simpler ways of assessing groundwater availability before resorting to experts and their complicated and costly methods. Surely if water witching has as much a chance of determining where groundwater is as the most sophisticated geophysical methods simpler ways must exist to give an indication of available groundwater resources. Is there any correct answer? So I decided to answer two questions. How does the different recharge estimation methods relate to each other and cannot a simpler way, based only on physical and rainfall properties, be found to give a preliminary estimate of groundwater recharge? The quest to answer this question is what follows in this dissertation.

But the road was fraught with roadblocks. My catchment had no data and a monitoring network had to be established. The logistics of such an approach chewing much valuable time from the real work. The equipment that was installed often failed at crucial periods such as the water gauge for the hydrological year 2000/2001. Finances were never adequate to put in place a comprehensive monitoring regime. Short cuts were taken and in some cases quality was compromised. Then there was my inexperience as a researcher and family catastrophe as two members of my family passed away. In the end the study was a journey against the odds and its conclusion more a triumph of determination over adversity than a scientific accomplishment. The output from this study is merely a beginning since the proposed model and its approach need refinement and to be tested in different regions as well.

SUMMARY

Hydrogeologists and Priests have one thing in common: they talk of things we cannot see but must believe in to live. Groundwater is vital, important and fascinating. Its occurrence has long been associated with sacredness. Groundwater recharge is perplexing and the recharge estimation methods are nothing more than an attempt by man to unmask this sacredness. While most methods give reasonable long-term annual average estimates very few if any at all offer guidance on short-term monthly recharge. In crystalline basement aquifers (CBAs) the problem is compounded by the high storage to flux ratio, which results in groundwater resources that are highly seasonal and high intra-annual and inter-annual variability.

This study proposes a simple method for assessing the potential available monthly and annual recharge in a small CBA catchment taking into consideration the rainfall and physical characteristics of the catchment. By studying the geology, land use and rainfall and applying different techniques for estimating recharge it has been deduced that occurrence of groundwater recharge, and hence the potential for recharge, is controlled by the rainfall regime and evaporative fluxes whilst the geology and vegetation characteristics of a catchment influence the mechanism and magnitude of actual recharge.

Three methods have been used to estimate groundwater recharge in a small catchment (180 Mm^2), the chloride mass balance (CMB), the daily catchment water balance (WB) and the water table fluctuation (WTF) methods. The study concluded that though the methods yielded the same range of recharge, between 8% and 15% of annual rainfall, all methods show a high spatial variability with coefficients of variation of up to 65% indicating that no single point measurements of recharge is a good indicator of regional recharge. The study also showed that all recharge estimation methods used in the study had the weakness of over reliance on one critical parameter such as the chloride deposition for the CBM method and specific yield for the WTF method. The WB method, at a daily time step, was influenced heavily by the spatial distribution of rainfall. The use of groundwater models such as MODFLOW was found to be of limited value in assessing recharge in CBAs due to the high heterogeneity in the geological properties of the aquifer.

The study showed that renewable groundwater resources in the Nyundo catchment have a storage period in the order of 100-120 days (3-4 months) whilst daily potential interception can be as high as 5 mm/d. A combination of multiple regression and interception models suggest that it is feasible to express the monthly recharge estimate as a function of its influencing factors; rainfall (number of rain days, average monthly depth), daily interception threshold, transpiration impacts (root depth), lithological properties (effective porosity) and state of the groundwater system (depth to water table). On the basis of such an analysis a simple model for assessing potential recharge using rainfall characteristics and soil and vegetation properties has been proposed. As average parameter values are used in the assessments a quasi-regional estimate of the groundwater recharge is obtained.

Table of contents

CHAPTER ONE

1. INTRODUCTION

1.1 General

Groundwater is generally potable at source, is available in-situ and has a low temporal variability making it the most important source of water for rural communities. Groundwater development is however complicated by highly variable hydrogeological conditions rendering its management fraught with uncertainty (Taylor & Barrett, 1999). In crystalline basement aquifers (CBAs)[1] uncertainty is highest and the groundwater potential least.

Crystalline basement aquifers occur in many parts of the world. The basement complexes are commonly known as shields. These shields are composed mainly of metamorphic and magmatic rocks of Precambrian age. Major shields are located in Canada, Australia, Amazon, the Baltic, India and Africa. At global scale their areal extent is estimated at 20% of the present land surface (Gustafson & Krasny, 1994).

On the African continent crystalline basement complexes occur extensively. They cover between 35% and 40% of the total surface area of the continent distributed mainly in the west, east and southeastern parts of the continent (Batchelor et al., 1996). Fig. 1.1 shows the distribution of hard rock complexes in Africa.

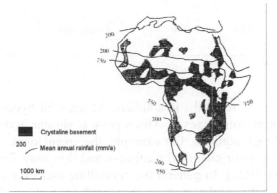

Fig. 1.1. The distribution of hard rock complexes and average annual rainfall in Africa. (Adopted from Batchelor et al., 1996)

Because of their poor groundwater potential coupled with the difficulty of studying them, the crystalline basement aquifers have not received much attention over the years. But due to the rise in water demand for potable water and the risk of contamination there has been a proliferation of studies on these aquifer systems (Taylor & Barrett, 1999). The improvement in standard of living with its attendant

---

[1] Aquifers underlain by an impermeable rock mass.

increase in the use of productive[2] water will add a further strain on the groundwater resources of these aquifers.

1.1.1 Physical characteristics of crystalline basement aquifers

In general there are three types of aquifers namely: crystalline aquifers, consolidated aquifers and the unconsolidated aquifers. The aquifers are derived from crystalline, sedimentary/meta-sedimentary and unconsolidated sedimentary rock formations respectively. An aquifer is characterized by two main intrinsic properties namely porosity and permeability, which affect the manner of storage and transmission of water (Wilson, 1969; Price, 1985). Hard rock aquifers possess secondary[3] or double porosity[4] and secondary permeability (Lloyd, 1999; Chilton et al. 1995), whereas unconsolidated aquifers possess primary porosity and primary permeability characteristics (Nonner, 1996).

Table 1.1. Some mean hydraulic properties of crystalline basement aquifers in Southern Africa. (adopted from Chilton & Foster, 1995).

Lithological unit	Saturated thickness (m)	Hydraulic conductivity (m/d)	Transmissivity (m^2/d)
Residual subsoil (Chimmbe, Malawi)	4	0.003	0.01
Regolith (Livulezi, Malawi)	17.6	0.31	5.5 (1-20)*
Regolith (Masvingo, Zimbabwe)	13.3	0.39	5.2 (1-60)
Regolith (Various)	12.9	0.36	4.6 (0.2-40)

* Values in brackets are ranges.

Table 1.1 presents some cases of regoliths in southern Africa with hydraulic details. The hard rocks of the basement complexes in Africa possess similar hydrogeological characteristics as other hard rock aquifers. For example, the greenstones[5] weather out quite easily and rapidly due to their chemical weakness and thus may form the main aquifer (Martinelli & Hubert, 1985). In general the crystalline rocks have little or no primary porosity and permeability, and as a result the search for features that enhance the negligible primary porosity is of utmost importance. Simple weathering processes, fractured fissure systems, and networks of joints and cracks in these crystalline rocks can develop secondary porosity (Wright, 1992; McFarlane, 1992). However the areal extent and depth of these rocks greatly determine the size of the aquifer. In undulating and rugged terrain, this phenomenon leads to localized and regionally disconnected aquifer systems (Wright, 1992; Chilton & Foster, 1995).

---

[2] Commercial or non-potable uses of water.
[3] Due to fracture crevices
[4] A combination of pore space and fracture crevices.
[5] A constituent rock type in basement complexes.

2

In addition to the general tectonic events the intrusion of dolerite dykes, quartz and pegmatite veins also cause fracturing in the rocks. The intrusion creates sufficient subsurface conduits to facilitate the movement and storage of groundwater (Pitts, 1984).

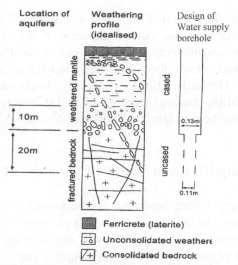

Fig. 1.2. Vertical representation of the weathered crystalline base aquifer system. (After Taylor & Howard, 2000).

In general a vertical profile of a weathered crystalline basement aquifer consists of a soil matrix at the top underlain by decomposed, fractured and fresh rock. Taylor & Howard (2000) and Chilton & Foster (1995) proposed an idealized weathered crystalline aquifer system illustrated in Fig. 1.2. The implication is that water storage is predominantly in the weathered mantle whilst the fractured zone serves as a transmission conduit within the aquifer (Gustafson & Krasny, 1994; Batchelor et al., 1996; Taylor & Howard, 2000). Naturally hydraulic conductivity in the fractured zone is as spatially variable as the nature of the faults and fractures within the rock such that borehole yields can differ by several orders of magnitude within the same rock unit and often within short distances (Gustafson & Krasny, 1994).

Some authors have proposed geological models for the crystalline basement aquifers focusing mainly on the decomposed/weathered zone, often referred to as the regolith. In hydraulic tests for crystalline basement aquifers in Malawi, McFarlane (1992) and Wright (1992) suggested a subdivision of the regolith into distinct units with different hydraulic properties and therefore different hydrological significance. The top most layer is composed of residual soil, is almost always in the vadose zone and is significant with respect to recharge potential. On the interfluves[6] the layer is visibly sandy with high infiltration capacity. Clay content increases with depth due to leaching and with it a decrease in porosity. The underlying layer, often referred to as saprolite, therefore has a high proportion of clay minerals, predominantly kaolinite, corresponding to the degree of weathering and leaching in the overlying layer. Water

---

[6] The higher altitude areas in a catchment.

transmission is likely to be lower than in the overlying soil zone. A third layer is made up of decomposed parent rock in which coarse-grained massive crystalline rocks, some without any detectable mineralogical change, still exist. Often this layer is called the saprock. The coarse-grained nature is often associated with the presence of further opened up fractures and therefore a high water potential.

Other important characteristics are the saturated thickness as this would determine the extent and rate of weathering and thus the storage capacity of the regolith (Pitts, 1984). Under prolonged saturation weathering is less aggressive resulting in coarse, partly weathered sand sized clasts which form the main aquifer in such systems (McFarlane, 1992; Taylor & Howard, 2000). Generally horizontal hydraulic conductivities are low in the regolith, rarely exceeding 0.1 m/d (Chilton et al., 1995), and negligible in the fractured zone below, ranging between 10^{-6} and 10^{-8} m/d (Taylor & Howard, 2000).

1.1.2 Crystalline basement aquifers as a source of groundwater resources

The importance of the African crystalline basement aquifers with respect to water resources varies from place to place. In the humid areas, where water is abundant, the interest in them has been to resolve water quality and engineering construction problems. In semi-arid and arid regions potable water supplies may only be from these aquifers as surface water streams are usually ephemeral or non-existent. In the arid regions studies have naturally sought to enhance the usefulness of the crystalline basement aquifers for water supply (Wright, 1992; Gustafson & Krasny, 1994; Chilton & Foster, 1995; Taylor & Howard, 2000). The studies have concluded that the importance of crystalline basement aquifers is that they are wide spread in areas of relatively high rural population densities, their groundwater is sufficiently shallow to allow low cost exploitation methods and that surface sources may be unreliable in these areas.

However, because of their limited transmissivity, crystalline basement aquifers exhibit comparatively low yields. Chilton & Foster (1995) report transmissivities of between 1 and 5 m^2/d with yield ranges generally less than 86 m^3/d. Such yields are considered adequate for most rural community borehole water supplies with demands of between 1-3 mm/a.

Knowing the demand imposed on the aquifer and the yield it can offer in meeting that demand is only part of the problem. A more crucial requirement is understanding how the aquifer is replenished and how sustainable the relation between this replenishment or recharge and the yield can be. This relationship is important particularly in times of drought to guide where to site wells (Wright, 1992) and how to manage the ground and surface water resources conjunctively.

Groundwater occurrence in the crystalline basement aquifers is characterized by the presence of a shallow water table and recharge is mainly from rainfall (Sekhar et al., 1994). The water levels have been observed to follow a seasonal fluctuation pattern influenced by the rainfall pattern. Several methods, mainly the chloride mass balance (Edmunds et al., 1988; Sangwe, 2001; Beekman & Sunguro, 2002) and baseflow separation (Farquharson & Bullock, 1992) have been used to estimate the groundwater recharge in crystalline basement rocks of Africa but none of them, separately, gives a satisfactory result because of heterogeneity and discontinuity of the aquifer and the complex nature of the resultant flow system (Chilton & Foster, 1995). However recharge rates of 0-25% of annual rainfall are often quoted in these aquifers.

4

Such estimates tend to be from very localized data leaving the problem of regional estimates unresolved (De Vries & Simmers, 2002).

1.2 Problem identification & research objectives

From the preceding summary a number of issues can be identified.

The first is physical and scientific, namely can we estimate, in a cost effective way, the amount of groundwater recharge to a crystalline basement aquifer system, at a spatial regional scale of say, a catchment? The second problem is management, can the recharge estimate be made over a more managerially meaningful time scale, say a month, so as to optimize the use of the water resources in the crystalline basement aquifer system? Given the shallow nature of the groundwater levels and the fact that water supplies by and for rural communities are from hand dug wells and therefore hardly are sunk below ten metre depth should the effort then not be directed to understanding the water dynamics of the regolith rather than the deeper fractured zones? Worse observations, by Davies & Turk, 1969 (cited by Lloyd, 1999) actually suggest that the marginal yield of water decreases with increasing depth for wells in hard rock aquifers.

The available literature seems to suggest that most authors see the problem of water availability in the crystalline basement aquifers as a geological problem. The shallow nature of the system however, means that the problem is also hydrological in nature and the aquifers need to be studied from a hydrological cycle perspective if optimal use of the water resources is to be achieved. The same literature shows that the recharge problem has extensively been attacked as a scientific problem and not a managerial one. As such the results, though scientifically sound, do not always address the actual managerial problem they are intended to solve. For example quoting long-term recharge in an aquifer that has a residence time for groundwater of about three to six months, as is the case with the shallow crystalline basement aquifer systems is somewhat superfluous.

Viewing groundwater recharge as a purely managerial problem, on the other hand, has resulted in undue focus on the exploitation of the resource as opposed to understanding the dynamics of the hydrogeological conditions that govern it. Over-exploitation or over-investment has often been the result.

1.2.1 Problem statement

Can a cost-effective rapid assessment method for estimating regional recharge, focusing on the regolith mainly, be found that both harnesses the scientific advances in understanding recharge and at the same time answering immediate managerial concerns?

1.2.2 Research objectives

Two basic objectives summarize the importance of this study.

To investigate the applicability of the common methods used to estimate groundwater recharge in a crystalline basement aquifer.

To develop a 'hybrid' methodology for estimating recharge in crystalline basement aquifers using simple hydrological and physical data.

1.3 Research methodology and limitations

The research methodology is based on application of the water balance at different spatial and temporal scales. Water balances were made for the whole catchment and for the saturated zone. The unsaturated zone was sparingly considered. Though some data was collected at hourly and daily time steps the main analysis was for monthly, seasonal and annual time steps. The spatial scales ranged from a local point to the catchment level (180 Mm^2). The main components of the water balance were measured directly in the field. The key measurements were groundwater levels, rainfall and catchment outlet discharge for hydrological processes, soil properties and lithological distributions (vertical and horizontal) for aquifer descriptions.

To investigate recharge as the most crucial component of the water balance, methods commonly applied to estimate recharge have been employed. These are namely, the water balance, the water table fluctuation, the chloride mass balance, the baseflow separation technique and groundwater flow modeling using MODFLOW. The results from each enabled comparisons to be made and the latent features of the methods as they apply to groundwater recharge in crystalline basement aquifers to be investigated in detail.

1.4 Thesis outline

Fig. 1.3 summarizes the study approach and thesis outline. The thesis is divided into nine chapters grouped in three parts. The first part defines the underlying principles and concepts of the study and consists of chapters one, two and three. Chapter One outlines the framework of the study. Chapter Two reviews groundwater recharge estimation and groundwater management. Chapter Three discusses the water balance and recharge estimation techniques.

The second part aims to characterize the study area and comprises of chapters four and five. Chapter Four is an introduction to the study area; its physical characteristics, water resources and human settlement and influence. Chapter Five describes the procedures followed in the study. It describes the fieldwork undertaken, analyses the data collected and summarizes the general characteristics of the catchment in relation to hydrology and hydrogeology.

Section three focuses on the analytical procedures and findings. The section consists of four chapters, six to nine. Chapter Six describes the water balance of the study area whilst Chapter Seven discusses the groundwater recharge estimation methods and the results from the catchment. Chapter Eight synthesizes the findings of the study whilst Chapter Nine gives the conclusions and recommendations from the study.

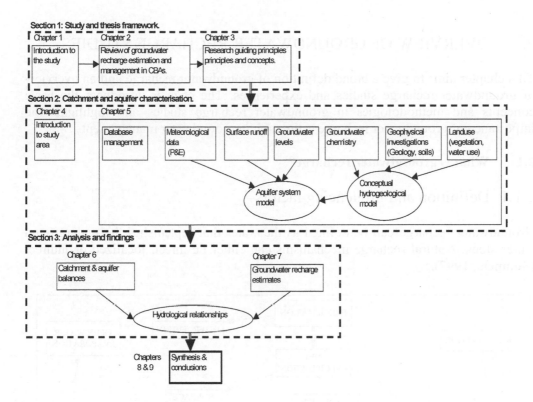

Fig. 1.3. Study approach and thesis layout.

CHAPTER TWO

2.   OVERVIEW OF GROUNDWATER RECHARGE STUDIES

This chapter aims to give a broad definition of groundwater recharge and an overview of groundwater recharge studies and experiences. The chapter introduces the basic concepts and methodologies in groundwater recharge studies and highlights the importance of groundwater recharge estimation in groundwater management.

2.1   What is groundwater recharge?

2.1.1 Definition and influencing factors

Groundwater recharge can be defined as the water that replenishes the underground water stock. Natural recharge mechanisms can either be direct, localized or indirect (Simmers, 1997).

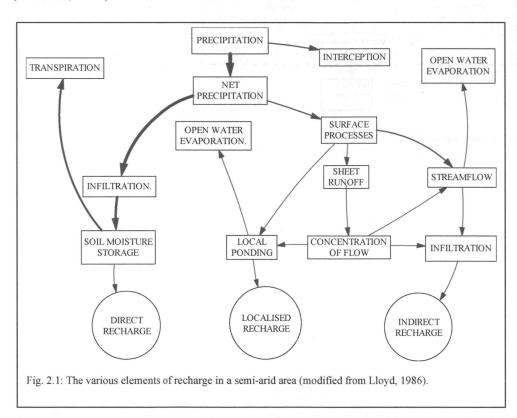

Fig. 2.1: The various elements of recharge in a semi-arid area (modified from Lloyd, 1986).

Fig. 2.1 illustrates the three types of recharge. Direct recharge refers to replenishment of the ground water reservoir from precipitation after subtracting interception, runoff and transpiration. Localized recharge results from the ponding of surface water in the absence of well-defined channels of flow whilst indirect recharge refers to percolation to the water table from surface watercourses. Each type of mechanism is more prevalent in some climatic conditions than others (Lloyd, 1986).

The accuracy of groundwater recharge estimates depends to a great extent on the correct identification of the recharge mechanism. However, no actual recharge process is strictly as defined above. Actual recharge is likely to be a combination of any of the above processes (Lerner et al., 1990). In arid areas a further problem is that when the basic recharge mechanisms are reasonably defined, deficiencies still remain in quantifying the elements particularly in areas of data scarcity (Lloyd, 1986).

Rushton (1988) identified several factors that affect natural groundwater recharge. At the land surface, recharge is affected by the topography and land cover in addition to the magnitude, intensity duration and spatial distribution of precipitation. Steep terrain supports fast fluxes and therefore favours surface runoff in preference to infiltration. A similar effect is achieved by a rainfall event of high intensity and low duration in the absence of preferential flow pathways to the saturated zone.

Vegetation cover influences the recharge process. For example, Sandström (1995) observed that baseflow decreased after deforestation in the Babati District of Tanzania and argues that this may have resulted from a combination of reduced preferential flow along roots and the bare ground that favours runoff.

In the soil, its nature, depth, cohesion, homogeneity and hydraulic properties determine recharge whilst the aquifer porosity determines the magnitude of recharge.

It can be concluded that the hydro climate, soil texture and slope play a role in determining the baseflow (and indirectly recharge) from a catchment.

2.1.2 The concept of hydrogeological provinces

Lerner et al. (1990) stipulate that before a quantitative evaluation of recharge can be made a conceptual hydrogeological model has to be built. The accuracy of the recharge estimate depends largely on the correlation of the conceptual model to the actual physical model. A region to be investigated is characterized as a hydrogeological province on the basis of climate, geology and soil type. Of relevance to this study are the mountain fronts, sand and the plutonic crystalline provinces.

The mountain front province is prevalent in arid to semi-arid regions of the world. Distinguishing features are that (1) recharge is by infiltration through gravels of high permeability and (2) the recharge water is collected from a larger area than the recharge surface itself.

Fig. 2.2 shows a cross section of a typical mountain front province. De Ridder (1972) divided the mountain front recharge into three hydrological zone, a recharge zone dominated by vertical downward flow, a transmission zone characterized by horizontal flow and a discharge zone where intersection of the piezometric surface with the topographic surface occurs.

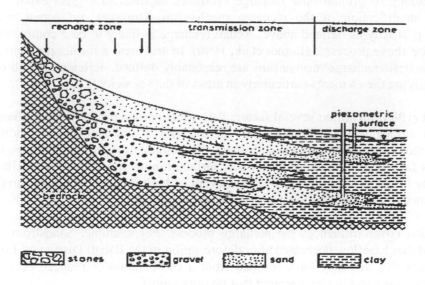

Fig. 2.2. Cross-section of an alluvial plain (After De Ridder, 1972).

Wilson (1980) showed that recharge is positively correlated to the size of the contributing catchment and that the recharge results from (1) infiltration from streams draining the mountains or (2) subsurface flows from the mountain mass itself. He further observed that the amount and rate of stream discharge are functions of the precipitation and the permeability of bedrock in the mountain mass whilst the occurrence and distribution of recharge from subsurface flows is affected by the topography and stratigraphy of the mountain mass.

Examples of sand provinces include the Sahara, Arabian and Kalahari sands. Sand is a primary product from the weathering of crystalline rocks or the secondary product of the decomposition of sandstone. The looseness of sand leads to high particle mobility, restrained vegetative growth as well as direct rapid recharge.

Lloyd et al. (1987) explained that whilst the in-situ moisture content in these provinces is a function of climate and grain size distribution, the groundwater recharge has a direct but non-linear relationship with rainfall, particularly the storm regime.

Recharge studies have reported natural recharge rates in sand provinces of between 3% and 21% (Sharma & Gupta, 1985; Caro & Eagelson, 1981). However, Allison & Hughes (1983) have suggested that macro pores along living roots or in dead root channels contributed to recharge in vegetated areas of the sand provinces.

In crystalline plutonic provinces groundwater recharge is dependent on the mode of chemical weathering and the rate of fracturing in the basement material (Lerner et al., 1990). Chemical weathering may produce layers of clay that result in perched water tables and limited groundwater recharge in deeper, more permeable formations. Mineralogy determines the faulting type and hence the amount of water that can be trapped and the directions in which it can flow. A high quartz content for example, results in brittle material exhibiting long fractures. Studies have shown that groundwater recharge rarely exceeds 15% of annual rainfall and that the salinity and isotopic content of the groundwater is usually the same as that in rainfall except in deep fractures where mineral dissolution comes into play.

Sukhija and Rao (1983) suggested that in granitic terrain the recharge is dependent on the annual rainfall as well as the potential evaporation. Recharge studies have quoted estimates ranging between 5% and 10% of annual rainfall. Athavale (1985) and Muralidharan et al. (1988) used a combination of the leaky aquifer modeling concept and tritium profiles to determine recharge rates of between 5% and 8% for an annual rainfall of 650-750 mm/a in the Vendavati river basin in India. Thiery (1988) used a lumped model approach in Ougadougou, Burkina Faso and obtained similar values. Houston (1982) used a combination of baseflow separation, hydro chemical analysis and simulation modeling to obtain similar results in Zambia and Zimbabwe. However for an area with less than 300 mm/a annual rainfall and high potential evaporation Allen and Davidson (1982) found very low recharge rates of 0.05% to 0.5% of annual rainfall.

The hydrogeological provinces for natural terrain are seldom as clear cut and distinct as described above. Most terrain is a combination of the above provinces. The upper parts of a given catchment can be predominantly plutonic whilst the lower parts are alluvial. For example, in Malawi the areas in the Rift Valley system are predominantly plutonic whilst those around Lake Malawi are generally alluvial.

The Nyundo catchment exhibits the characteristics of a mixed province. The climate is semi-arid and the underlying basement is of igneous origin. The upper parts of the catchment are dominated by decomposed and fractured granite whilst the valleys are predominantly composed of Karoo sand deposits with isolated granitic outcrops. Such a geologic configuration gives the catchment the characteristics of a sand province bordered by a crystalline plutonic province. (See also Section 5.3).

2.1.3 Groundwater recharge estimation methods

Most authors agree that the estimation of recharge is best carried out as an iterative process since data is always limited and circumstances vary both in space and time (Simmers, 1997; Lerner et al., 1990; Lewis & Walker, 2002). The first step is to define the groundwater system in terms of the geological structures and the resultant flow mechanisms. Second, the complete water balance must account for all water that 'does not become recharge' and the underlying groundwater recharge processes clear. Third, the estimate must consider the time scale for the recharge process. As a rule Lerner et al., (1990) state that estimates based on the summation of shorter time steps are better than those based on longer time steps for the same duration.

Lerner et al., (1990) identifies five methods used for estimating groundwater recharge; direct measurement, water balance techniques, Darcian approaches, tracer methods and empirical relationships.

Direct measurement of groundwater recharge is made in lysimeters installed at depth to limit the influence of surface processes including interception and surface runoff.

The advantages are that the method directly measures balance components and can cover a wide range of time steps from brief events to seasonal variations. A disadvantage is that only a point estimate is taken thereby limiting the applicability of the method to catchments with homogenous features.

The water balance approaches measure recharge as a residue of other fluxes that are easier to measure or estimate such as rainfall, evaporation and discharge. They include soil moisture balances and water table fluctuations evaluated over time scales of events to seasons. (See also Section 3.1).

11

The advantages of the balance methods are that they are based on measurable data, easy to apply and account for input water. The disadvantages are that recharge is estimated as the difference of comparably large values thereby increasing the relative error of estimate and that the independent variables are in themselves often difficult to measure accurately.

The Darcian approaches are based on the law of Darcy for flow in a saturated porous medium and have state variables (pressure heads, hydraulic conductivity, etc.,) and boundary flux conditions as input. Their main advantage is that they represent the actual flow conditions albeit with simplifying assumptions. One such necessary simplification is to have an adequate schematization of the geological formations, which due to the high heterogeneity of most flow media is difficult to attain. This drawback is illustrated in Section 7.4 of this thesis.

The flow can be analyzed in three dimensions but is often simplified into two dimensions or even one dimension depending on the simplicity of the case in point. Usually models are used to ease and speed up computations. MODFLOW and EARTH are examples of models that can be used in the saturated and unsaturated zones respectively.

The general equation for Darcian flow in three dimensions can be written as follows:

$$\frac{\partial}{\partial x}\left(k_x\,\frac{\partial \phi}{\partial x}\right) + \frac{\partial}{\partial y}\left(k_y\,\frac{\partial \phi}{\partial y}\right) + \frac{\partial}{\partial z}\left(k_z\,\frac{\partial \phi}{\partial z}\right) - S_y \cdot \left(\frac{\partial \phi}{\partial t}\right) = R \qquad (2.1)$$

where x, y and z are Cartesian co-ordinates, k [LT^{-1}] denotes the hydraulic conductivity, S_y [-] is the specific yield, t [T] is time, ϕ [L] is the hydraulic head above an arbitrary datum and R is the recharge into the system. As Equation 2.1 illustrates the Darcian approaches relate changes in storage to the movement characteristics of a dominant flux which limits their application to cases with measurable and describable flow.

Tracer methods rely on environmental and applied tracers to date, label and quantify recharge. Signature methods seek to label the recharge and are exclusively based on applied tracers like fluorescent dyes. Throughput methods on the other hand seek to quantify the recharge through the mass balance of the tracer. Conservative tracers like chloride are the norm in most investigations. Though convenient and sometimes the only available method for arid areas (Beekman et al., 2000) tracer methods suffer the handicap of mass distortions from secondary inputs and mixing and/or dual flow mechanisms. (See also Section 3.2). Generally, the tracer methods are useful for long time scales in the order of years.

Empirical methods seek to correlate recharge with other measurable hydrological data such as rainfall and surface flow through the use of mathematical formulae such as shown in equations 2.2 to 2.4 (Sami & Hughes, 1988; Mandel & Shiftan, 1981 cited in Lerner et al., 1990; Sinha & Sharma, 1988 cited in Lerner et al., 1990). In all cases R and P denote recharge and rainfall respectively whilst k and n are constants reflecting physical conditions.

$$R = k_1 \cdot P \qquad (2.2)$$

$$R = k_1 \cdot (P - k_2) \qquad (2.3)$$

$$R = k_1 \cdot \left(\frac{P}{k_2} - k_3 \right)^n \qquad\qquad (2.4)$$

The advantage of such approaches is that they can be transposed in time and space and render themselves practically useful for preliminary recharge estimates. Their main disadvantage is that they are site specific and derived from other methods of recharge estimation. As such they can only be as accurate as the methods from which they are derived. Secondly, since they are not physically based they can be rendered obsolete by changes in catchment physiology if not reviewed periodically. Despite this setback, the author considers them of valuable importance in estimating recharge for water resources management purposes particularly in areas of data scarcity and limited technical and financial resources.

An improvement to the empirical methods can be termed the 'hybrid method'. In this approach combining physically based techniques with empirical methods reduces somewhat the shortcomings of the latter. Such an approach combines the influence of climate, geology, terrain geometry and land cover on recharge into a single estimate. This approach is the focus of this discourse.

Some challenges remain in the estimation of groundwater recharge. De Vries & Simmers (2002) still identifies as challenges, the need to fully understand the temporal and spatial variability of recharge, development of methods to estimate areal recharge especially from point measurements and fully deciphering the impact of landuse particularly urban developments on groundwater recharge. The water resources management challenge however, is to find cost-effective, simple rapid assessment methods for estimating recharge at time steps that allow real-time water resources management decisions to be made.

2.2   Groundwater recharge studies - state of the art

2.2.1 Understanding groundwater recharge processes

It is generally accepted that the interaction of climate, geology, morphology, soil condition and vegetation determine to varying degrees the recharge process (Simmers, 1997, De Vries & Simmers; 2002). Groundwater recharge in semi-arid areas is more susceptible to near surface conditions as compared to humid areas. Further, the potential evaporation is higher than the rainfall making recharge dependent on rainfall intensity and the existence of fissures and cracks that favours preferential flow in the receiving soil mass.

However, more subtle considerations may affect the recharge process. Defining direct groundwater recharge as the downward flow of water to the saturated zone creates conceptual problems regarding the recharging processes (De Vries & Simmers, 2002). Just as the net rainfall reaching the ground is reduced by interception (De Groen, 2002), not all percolating water necessarily reaches the water table.

Percolation may be hampered by low vertical conductivity horizons resulting in lateral flow to nearby depressions. Similarly in shallow aquifer systems local seepage may result in net recharge being lower than the total addition to the water table. Direct evaporation and transpiration from the saturated zone has a similar effect (De Vries & Simmers, 2002).

Mechanisms also exist that may cause water ascend from considerably deep water tables especially in semi-arid to arid areas. Capillary rise, induced by deep rooted vegetation, has been reported to cause upward fluxes of up to 1 mm/a from depths ranging between 15 m and 50 m (Adar et al., 1995; Haase et al., 1996; De Vries et al., 2000).

Recently, vapour transport has been observed to produce considerable vertical fluxes in areas were a temperature gradient exists. De Vries (2000) reported a winter upward and summer downward vapour transport flux of 0.2 to 0.3 mm per season due to a temperature gradient of $4\ ^0C$ between the lower limit of the root zone at 3 m depth and the 7 m depth level.

2.2.2 Trends in groundwater recharge estimation methods

Lerner et al., (1990) and Simmers (1997) content that tracer methods of estimating recharge are more successful than the indirect physical methods because they are simple to use, relatively cheap and universally applicable. However, Cook and Walker (1995) caution that vapour transport particularly for tritium profiling, can affect the results for recharge estimates below 20 mm/a. Selaolo (1998) further warns that not only is it difficult to determine the atmospheric deposition of chloride but the deposition has considerable temporal and areal variability. Despite these shortcomings tracer methods remain the most widely used for all types of groundwater recharge estimates in semi-arid areas. To reduce the margin of error in estimation the trend has been to use methods in combination either as multiple tracer methods or as tracer and some physical method.

The application of Darcian methods to the unsaturated zone has been hampered by their reliance on soil parameters like the unsaturated hydraulic conductivity which are difficult to determine. Researchers such as Van Genuchten et al. (1992) have developed simplified methods based on soil data to counter this handicap. Effort has also been made to estimate groundwater recharge as a function of soil temperature changes in the unsaturated zone (Taniguchi & Sharma, 1993).

The effect of land use has been investigated using 'at point' Cl profiles at different times for the same point (Walker et al., 1991). Complementary plot size studies have been conducted to investigate the small scale variation in recharge. Jolmston (1987) obtained a variation of 2.2 to 100 mm/a over a plot area of 700 m^2. Butterworth et al. (1999a) showed that percolation rates differed for different soil types or the same soil type with different surface preparation for the same plot area.

Several studies have shown that reasonable regional groundwater recharge estimates can be obtained using readily available field data without considering small-scale local variations. Methods used include isotope dating (Adar et al., 1988), chloride mass balance calculation and mixing cell modeling (Gieske & De Vries, 1990), Darcian flow modeling (Van der Lee & Gehrels, 1997) and direct measurement of stream and/spring discharge (Kennett-Smith et al., 1994). Recently studies have explored the use of field measurements, remote sensing data and geostatistics. Sophocleous (1992) combined regression analysis of field data with GIS overlays to determine regional recharge whilst Kennett-Smith (1994) related recharge to soil type, rainfall and land use data by combining field data with a daily water balance. Salama et al. (1994) showed that recharge areas can be identified over a 2000 Mm^2 region using aerial photos and Landsat imagery.

14

Concern has been raised over the hitherto standard practice of expressing recharge as a percentage of annual rainfall in areas where recharge results from infrequent large rainfall events (Lewis & Walker, 2002). For example, De Vries & Simmers (2002) showed that a plot of recharge against annual rainfall in three countries, Zimbabwe, S. Africa and Botswana gives variations of up to 100 mm/a for the same rainfall amount.

2.2.3 Groundwater recharge studies in SADC[7]

In the Southern African Development Community (SADC) groundwater recharge estimation has been studied extensively in South Africa and Botswana and sporadically in Malawi and Zimbabwe. Studies in the other SADC countries are still embryonic.

Various methods and techniques ranging from tracer and stable isotope profiling techniques to groundwater balance methods and pore soil moisture profiling have been used in the study of groundwater recharge in Southern Africa.

Botswana is taken as a sand province of Kalahari sediments underlain by Karoo strata. Rainfall ranges between 200 and 600 mm/a. Recharge studies in the country have focused on both the short and long-term effects of precipitation on recharge particularly the temporal and spatial distribution of rainfall. Methods used included analysis of hydrographic records and groundwater dating with tritium (^{3}H), carbon-14 (^{14}C) and carbon-13 (^{13}C) (Verhagen et al., 1984; Mazor et al., 1977; Foster ct al., 1982). Selaolo, et al., (2000) used helium isotopes and abundances, chloride mass balance, groundwater flow and hydrochemical isotope modeling. Recharge rates of 1% to 5% have been determined. Some conclusions were that recharge water can be from several years old and originate out of the aquifer's spatial boundary.

Groundwater recharge estimation studies in South Africa (SA) have been conducted using different direct and indirect techniques and methodologies in the unsaturated and the saturated zones.

Groundwater recharge in the unsaturated zone has been assessed using lysimeter studies, tritium profiling, soil moisture balance and chloride profiling techniques. Bredenkamp, et al. (1995) noted that lysimeter results indicate an apparent annual threshold value of rainfall below which no recharge takes place. Tritium profiling yielded estimates that were generally less than those derived from water balance methods (Bredenkamp, et al., 1978). Estimates between 5% and 18% of average annual rainfall were obtained (Verhagen, et al., 1979; Bredenkamp, et al., 1978). The soil moisture balance method was found to be too dependent on the assumed equivalent soil moisture available to the vegetation (Bredenkamp, et al., 1995 citing Masren, 1980). In general, the recharge rates estimated through chloride profiling were found to correspond well with values obtained using other techniques (Bredenkamp, et al., 1995). The values ranged between 0.9% and 37% of annual rainfall.

---

[7] Southern Africa Development Community

Two main methods, saturated volume fluctuation (SVF) and cumulative rainfall departure (CRD), were used to estimate groundwater recharge in the zone of saturation in SA. Hydrograph analysis and spring flow analysis were used in areas where spring flow and discharge data was available.

Botha (1994) applied the saturated volume fluctuation (SVF) method to estimate the groundwater recharge in a dolomite aquifer covering 70 Mm^2. Using recorded water level, rainfall and spring flow data for the period 1960 to 1993 and computer simulated groundwater level data for the same period. Recharge rates of between 10-14% of average annual rainfall were estimated. Such values agreed closely with those obtained from the chloride profiling approach.

The greater part of Malawi is composed of crystalline metamorphic and igneous rocks referred to as Basement Complex with minor occurrences of Karoo sedimentary formations and volcanic rocks confined in the northern and southern extremities of the country. However, the most imposing structural feature is the Rift Valley in which Lakes Malawi and Malombe and the Shire River are located.

Two main aquifer formations can be distinguished in Malawi (1) the extensive but relatively low-yielding, weathered basement aquifers of the plateau, and (2) the higher-yielding alluvial aquifers of lake shore plains and the Shire valley.

Smith-Carrington and Chilton (1983) investigated the overall groundwater resources in Malawi. They quantified the groundwater resource and its occurrence, derived the aquifers' hydraulic properties, evaluated the groundwater use and potential, analysed the groundwater hydrochemistry and classified areas according to their potential yields.

Four techniques were used to assess the groundwater: stream hydrograph analysis, groundwater level fluctuations, flow nets and catchment water balance techniques. The researchers suggested a general range of recharge of about 1-5% of the rainfall for the weathered basement aquifer and 1-7% for the alluvial aquifer.

Groundwater recharge in Zimbabwe has been studied on the Lomagundi, Save and Nyamandhlovu aquifers. Of late several attempts have been made to estimate the groundwater recharge in the crystalline basement aquifers of Zimbabwe. The crystalline basement complex covers approximately 60% of the country.

Table 2.1. Groundwater recharge estimates by studies done in Zimbabwe.

Aquifer type	Area	Estimated recharge (mm/a)	% of average rainfall	Estimation method	Author
Crystalline basement	Nyatsime catchment	131	16.4	Water Table Fluctuations	Mudzingwa & Lubezynski (1999)
		130	16.3	Chloride Mass Balance	
		74	9.3	Reservoir method	
		162	18.0	Flux analysis	
Crystalline basement	Marondera Grasslands	136	15.1	Chloride Mass Balance	McCartney (1998)
	Research Catchment	185	20.5	Chloride Mass Balance	Jarawaza (1999)
		190	22.0	Water Table Fluctuations	
Crystalline basement	Chiweshe, Mazowe	71.6	8.5	Chloride Mass Balance	Mjanja (2000)
		34.5-66.9	4.1-7.9	Baseflow Separation	
Dolomite	Lomagundi	100-150	12-18	Water Table Fluctuations	Wagner et al. (1987)

Adapted from Sankwe (2001).

Table 2.1 summarizes some of the studies in CBAs in Zimbabwe. Recharge rates between 4% and 22% of annual rainfall have been obtained. Different techniques do not lead to fundamentally different estimates suggesting little dependency of the estimates on the techniques employed.

2.3   Groundwater recharge and groundwater management

2.3.1 Groundwater management

Custodio (2002) noted that groundwater is a key factor to many hydrological processes which sustain spring discharges and river baseflow, lakes and wetlands, transport dissolved chemicals, facilitate weathering as well as provide habitat to micro organisms. The importance of groundwater does not end with eco-hydrological functions. Its greater importance is in sustaining human life.

Groundwater provides a high proportion of withdrawals amounting to more than 20% in most countries mainly for domestic supplies, mines, industries and irrigation. The rates of groundwater usage are high and increasing, e.g., in Egypt a 200% increase was recorded from 1975 to 1985. In the crystalline basement aquifers in general 30% of rural population has access to clean water abstracted mainly from dug wells with yields rarely above 430 m^3/d and median yields of 26 m^3/d (Lloyd, 1994). Despite its widespread use however, Llamas (1992) noted that though users, planners and developers appreciated the importance of groundwater as a source of good quality water, its economic advantages over surface water are still to be fully emphasized.

The main advantages of groundwater are that it occurs in-situ and its development is not dependent on large-scale collective projects requiring engineering structures for diverting, regulating and transporting the water. Thus the capital costs for groundwater are measurably lower than those for surface water which factor makes groundwater more ideal for poverty alleviation than surface water. Further, pumping costs are offset by low treatment costs when compared with surface water schemes.

From a management perspective groundwater has a number of weaknesses. As with surface water, the extraction of groundwater changes the water budget (Llamas, 1992). Second, water-using activities affect the quality of groundwater resources. The use of agricultural chemicals by farmers, sewage disposal and leakages and inadvertent dumping by industry pose a serious threat to shallow aquifers. Quarrying and mining activities require subsurface channels and dump sites that may be the conduit and storage for hazardous materials that threaten groundwater at depth. Once polluted groundwater may be too expensive or impossible to recover.

What is also worrying is that users very often are not aware of their transgressions and co-dependence and have a partial understanding of the groundwater resource. Partly because of the latter reason, groundwater resources are either under-appreciated and under utilised or inappropriately exploited and over utilised. Either way there is a dearth of sound groundwater management. Llamas (1992) argues that this poor management of groundwater resources may be in part due to the lack of basic knowledge among planners and decision makers on the origin, movement and pollution of the groundwater. It follows from this argument that if the dynamics of groundwater are better understood then management of the resource would most likely be improved. Such is the import of this study.

2.3.2 Groundwater yield, sustainability concepts and management policies.

The idea of groundwater recharge is a relatively young concept in the field of hydrogeology. Early men, and some wild animals to this day e.g. elephants, only knew that water can be obtained from the ground. Often this source of water, especially springs, attained a spiritual significance. Over the years groundwater came to be regarded as a mineral giving it a private status often enshrined in law (Water Act (Zimbabwe), 1976; Llamas, 1992; Van Tonder, 1999). During this period, the pre-occupation in hydrogeology was determination of yield defined broadly as the volume of water that can be abstracted from an aquifer per year. Naturally this approach led to mining of groundwater as demand increased and abstraction technologies improved. When the negative consequences of groundwater mining (increased pumping costs, land subsidence and reduced baseflow) surfaced new concepts to guide groundwater exploitation emerged.

The concept of *safe yield* was introduced by Meinzer in the 1920s. The safe yield is commonly defined as the groundwater abstraction that ensures the *"attainment and maintenance of a long term balance between the amount of groundwater withdrawn annually and the annual amount of recharge"* (Sophocleous, 1997). The concept enjoyed popular support up to the 1980s. In the USA, it was used to issue groundwater rights (Sophocleous, 2000). In Zimbabwe it was used to guide the issuing of groundwater permits (Martinelli, 2000).

The safe yield concept ignores the fact that under natural conditions recharge is balanced by discharge to streams, springs, wetlands and other natural outlets hence a safe yield consumption can dry up rivers and wetlands. These negative impacts of a safe yield policy take long to be noticed (Sophocleous, 2000; Custodio, 2002). As such the concept is erroneous and leads to unsustainable use of groundwater (Bredehoeft, 1997; Sophocleous, 1997).

The concept of sustainability introduced in the 1980s (Custodio, 2002) was a natural successor to the increasingly questionable safe yield approach. By definition, *"Sustainable development of a natural resource is the development that meets the needs of the present without compromising the ability of future generations to meet their needs"* (WCED, 1987). As Sophocleous (2000) points out sustainability is an idea broadly used but perhaps not well understood. Given the different levels of development in different countries or different locations within countries, sustainability may mean very different things from place to place. The concept remains the buzzword for groundwater resources management in particular and natural resources in general.

The concepts of the safe yield and sustainability are relatively easy to understand when applied to classical aquifers, i.e., unconsolidated deposits of measurable areal extent and depth with meaningful transmissivities and vast storages. When applied to crystalline basement aquifers some points need consideration.

The low storage potential of the crystalline basement aquifers, due to a near surface crystalline base and resultant thin regolith, mean that the rate of dry holes, defined as sited wells that fail to strike water, are as high as 10%-50% (Wright, 1989). Groundwater mining is not an option (Houston, 1992). As such a falling water level in crystalline basement aquifers signals not only an increase in energy needs to pump the water from greater depth but a real likelihood of groundwater exhaustion.

Gustafson & Krasny (1994) observed that the groundwater resources of hard rock aquifers strongly depend on present recharge capability. Therefore, groundwater management in these aquifers also depends on how well the seasonal recharge is assessed (Lloyd, 1994). Safe yield in this case may mean the abstraction rate that does not lead to complete depletion of the groundwater resource but may still affect baseflow. On the other hand the fact that the recharge is seasonal means sustainability concepts have no immediate physical meaning. What is more important is how to prepare for periods of reduced recharge such as drought years.

Management policies and legislation have tended to evolve not only in tandem with the hydrogeological concepts of the time but also in line with government ideologies of the day. The evolution of groundwater management in South Africa illustrates this case. In the past, when groundwater was regarded as a mineral and Apartheid was the ideology of choice for government, scant attention was given to groundwater resources development for rural communities. As a result groundwater was understudied, under-utilised and viewed disdainfully as a resource to be discarded as soon as more reliable surface water resources were identified (Van Tonder, 1999). With the change in government in the early 1990s and a general realization that surface water resources were already strained emphasis shifted to groundwater resources development. The South African Water Act (1998) stressed the need to protect, conserve and manage groundwater resources sustainably. In this regard a groundwater "Reserve" was defined for potable and environmental use and given the highest priority of all water uses. Similar changes have occurred in the SADC region and elsewhere.

The groundwater management problems in developing countries however go beyond hydrogeology and legislation. Lloyd (1994) points out that a mixture of legislation, national and/or external funding policies, fragmented local organizational structures and limited national infrastructure all contribute to hinder sound groundwater management. Development is further stalled by lack of long-term sustainable groundwater resources and the cost of obtaining alternative supplies. Data scarcity due in part to the absence of adequate hydrological monitoring networks or lack of finance to maintain existing infrastructure and conduct resource assessment studies result in constrained, if not inappropriate, management options. Thus, when groundwater has to be developed past usage is inconsistently known and when known records are too short to allow determination of significant trends (Llamas, 1992).

An important observation is that groundwater managers often lack the technical instruments analogous to surface water since groundwater is largely a hidden resource and its control 'more an art than a science'. To compound the situation further water managers tend to emphasize the surface water aspects of hydrology. Because of this monitoring and sound aquifer knowledge and calculation or modeling behaviour are needed to develop sound groundwater management objectives and policies (Custodio, 2002). A combination of legislative measures, management re-organisation and improved assessment methodology can ensure better management of groundwater resources. Needless to say future strategies need to be radically changed from current practices (Lloyd, 1994). This study seeks to contribute to this future by focusing on improved assessment methodologies.

While these problems are general with respect to groundwater management in third world countries in the arid and semi-arid crystalline basement aquifers added, if not prime, constraints to sustainable development are the low and erratic rainfall and limited availability of both the groundwater and surface water (Batchelor, et al., 1996). To address this problem hydrological and hydrogeological assessment methods have been improving over the years.

With respect to groundwater there has been a proliferation of groundwater recharge studies over the last two decades. Most of these studies have sought to answer the question of how much water enters the aquifer in a year. The more managerial question of how much of this water man can use has received less attention. Depending on the level of exploitation and the nature of the aquifer system, seasonal groundwater storage may be of greater importance than the average annual rate of aquifer replenishment. This is certainly the case with the crystalline basement aquifers. Given their shallow nature, subsequent dependence on direct rainfall recharge and the dominance of outflow fluxes the crucial management task is to estimate how much groundwater is available at certain periods of any given hydrological year rather than how much water enters the aquifer during the year. Groundwater recharge studies, dominated as they are by the assumption of the aquifer as sink storage tend to overlook this aspect of annual hydrological dynamics in the crystalline basement aquifers. In the classical aquifer storage-to-flux is relatively large (in the order of several years). For such a case the accuracy of the recharge estimate at small time steps may not be crucial and quoting recharge in terms of annual rates makes sense. In the crystalline basement aquifers where the storage-to-flux ratio is relatively low (in the order of days even) the accuracy of the recharge estimate at smaller time scales is important.

CHAPTER THREE

3.   THE WATER BALANCE AND GROUNDWATER RECHARGE ESTIMATION

This chapter explains the water balance theory as understood and applied in this discourse. The underlying ideas behind the water balance approach and its application to groundwater recharge estimation are discussed. Assumptions underlying the approach and its application to recharge estimation are presented and evaluated.

3.1   The water balance approach

3.1.1 The basic concept

Definitions

The water balance approach is defined as the application of the law of the conservation of mass to hydrology. The approach can be stated thus:

> *For a definitive volume of water during a specific time period, the difference between the total input into and total output from the volume is balanced by the change in storage in the volume.*

$$I - O = \frac{\mathrm{d}S}{\mathrm{d}t} \qquad\qquad (3.1)$$

Where I [LT^{-1}] denotes inflows, O [LT^{-1}] outflows, dS [L] the change in storage and dt [T] is the time increment. In other words the change in the stock over time equals the quantitative difference between the incoming and the outgoing fluxes.

To apply the water balance, stocks and fluxes, as defined by specific spatial and temporal boundaries, must be defined. Only the stocks within the boundary and the fluxes cutting across the spatial boundary but within the temporal boundary are considered. In a hydrological catchment, inflows usually consist of precipitation, upstream surface flows and lateral subsurface flows whilst the outflows consist of surface discharges, lateral subsurface flows and total evaporation. The changes in storage occur in host media such as the open water bodies (rivers and lakes), the unsaturated zone and the saturated zone.

Equation 3.1 implies that three scenarios can occur with respect to the water balance of a catchment. If the inflow is greater than the outflow catchment storage increases, if the reverse occurs then catchment storage decreases. The third scenario occurs when inflows equal outflows and the change in storage is correspondingly equal to zero. To avoid the difficulties associated with measuring changes in storage the third case is always utilized in hydrological studies by selecting a period in a time series over which the change in storage is zero, or by taking a long enough interval such that the storage fluctuation is small compared to the aggregated fluxes. Equation 3.1 can be written in its summation form as:

$$S(t) - S(0) = \int_0^{\Delta t} I dt - \int_0^{\Delta t} O dt \tag{3.2}$$

where I and O denote average values over Δt. If a long enough period of time, Δt, is chosen such that the left hand side of Equation 3.2 is made as small as possible and after division by Δt, it reduces practically to zero relative to the individual terms on the right hand side then the storage variation can be considered negligible yielding:

$$\Delta S = \int_0^{\Delta t} I dt - \int_0^{\Delta t} O dt \approx 0 \tag{3.3}$$

The terms to be included in a water balance equation reflect the complexity of the water balance and are an indication of the factors governing the hydrological process under consideration.

The selected boundary can be natural or artificial. The water balance components of a natural water balance like a catchment can be reasonably straightforward whereas those for artificial boundaries may be more complicated. The duration, Δt, in the water balance influences the number of components to be considered in the balance. Generally a longer duration leads to fewer components since rapid fluctuation processes cancel out by averaging over longer periods of time as shown above. The purpose of study and hence the available resources can also limit the number of components to be considered. If less accuracy is required as for example in water supply studies, fewer components can be used whereas detailed research studies require the evaluation of more components.

The hydrological regime for which the balance is applied also determines the number of components to be considered. If the aim is to evaluate low flows, studies can be conducted in a longer period in the dry season when some components like rainfall are negligible whereas the impact of floods can only be evaluated during the duration of the flood. Similarly groundwater recharge due to high intensity, short duration rainfall events in the arid regions may best be estimated on an event time scale or short daily intervals. Finally, data availability may turn out to be the overriding factor in determining the components of the water balance. Less data leads to a simplified balance equation.

Effect of temporal and spatial scales

Two types of water balance can be distinguished with respect to time. The mean water balance is applied over annual cycles, usually hydrological years defined as the period after which the hydrological events start to repeat themselves. Over a hydrological year the change in storage is generally small compared to the aggregated fluxes and can be equated to zero for computational purposes. The instantaneous water balance is applied for a short time interval, usually much shorter than a year. The storage change is therefore not always equal to zero and the need to disaggregate it into its constituencies arises. In the case of a river basin the total storage change can be written as:

$$\frac{dS}{dt} = \frac{dS_{ow}}{dt} + \frac{dS_s}{dt} + \frac{dS_u}{dt} + \frac{dS_g}{dt} \tag{3.4}$$

22

Subscripts s, ow, u and g denote ground surface, open water bodies, the unsaturated zone and the saturated zone, respectively. S and t denote storage and time respectively. Different terms can be used depending on the selected boundary and the storage components therein.

Time variability is important in considering the water balance as it gives an indication of the duration of the availability of a given stock. The time that a given stock requires to deplete under the influence of its natural discharge is the characteristic 'residence time' of the water in that stock.

Three time variations can be considered: the time step or interval, the incremental time and the time scale. The time step, Δt, is the time specified for computational purposes and can vary from an hour, a day, a month, a year or several years depending on the purpose of the calculation. It is the time over which the balance is considered and thus defines the absolute time boundaries. The incremental time, dt, is the infinitesimally small time step in a derivative. The time scale, T, represents the residence time as defined above. It is therefore a hydrological parameter and a property of the system. Mathematically it can be defined as:

$$T = \frac{S}{\left(\dfrac{dS}{dt}\right)} \tag{3.5}$$

where S and dS/dt represent absolute storage and the rate of change of that storage over time respectively. The time scale is directly proportional to the water stock but inversely proportional to the flux depleting it.

Combining Equations 3.1 and 3.5 gives:

$$T = \frac{S}{I - O} \tag{3.6}$$

Equation 3.6 suggests that T is the time scale of the stock variation, i.e., the change in stock over time with respect to all fluxes. This time scale is infinite if the influxes equal the outflows. This implies that the stock remains constant in time.

What is more relevant to hydrological analysis however, is the time scale of the individual processes. This is the time scale with respect to a particular flux and in practical terms is a measure of 'how long a particular water drop related to a flux stays in storage'. Viewed this way, it means at a certain point in time the time scale of the inflow into storage can be different from the time scale of the outflow from the same storage. Hence, there is an attenuation effect.

The time scale for the inflow can be written as:

$$T_I = \frac{S}{I} \tag{3.7}$$

And that for the outflow as:

$$T_o = \frac{S}{O} \tag{3.8}$$

In the long run, the time scales are equal since the change in storage will approach zero, and hence $I = O$. The process time scales T_I and T_o are the average residence times for the water in stock.

An example can be given here of the physical meaning of the process time scale with respect to stock depletion over a period where recharge is zero ($I = O$). In the case of catchment groundwater storage, natural depletion follows a first order decay given by:

$$S_t = S_0 \cdot e^{\frac{-t}{K}} \qquad (3.9)$$

which when differentiated and rearranged yields:

$$K = -\frac{S\,\mathrm{d}t}{\mathrm{d}S} \qquad (3.10)$$

It follows from Equations 3.5 and 3.10 that K [T] is the time scale for groundwater depletion, in other words, the residence time of groundwater in the catchment.

The magnitude of the time scale depends on the ratio between the flux and the stock rather than their absolute values. Thus for groundwater storage, which is large relative to groundwater discharge, the residence time is large. Surface water, in the case of no impoundment, has a small stock relative to surface flow and results in short residence times.

Errors and units

Since the magnitudes of the components of the water balance can never be accurately determined an error term needs to be incorporated into Equation 3.1.

$$I - O \pm \delta = \frac{\mathrm{d}S}{\mathrm{d}t} \qquad (3.11)$$

The error term accounts for measurement mistakes, computational weaknesses and/or inadequacy of techniques and conceptual failures in identifying or defining balance components with respect to temporal and spatial variations. Each component of the water balance has an error and all the individual component errors are accumulated in the final error term. Since the error term is a residual term of the water balance that sums up all the errors associated with the water balance if its value approaches zero it does not necessarily follow that measurements are accurate – it merely suggests that the error terms balance out. In cases where a component is calculated as the residual of other terms in a water balance equation then the error term is incorporated into the calculated value and ceases to be an independent variable.

The error term can be reduced if independent methods are used to quantify the components of the water balance.

All components of the water balance must be in the same units lest the arithmetic is meaningless. Usually balance terms are expressed in water depths or volumes per unit of time and the water balance is always taken over a specific period. Dimensional analysis of a balance equation is a basic requirement in water balance computations.

3.1.2 The water balance of a crystalline basement aquifer catchment

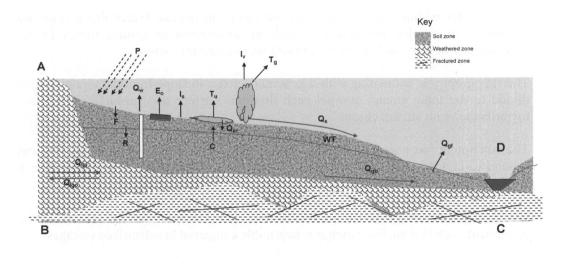

Key terms

Parameters for the surface & water bodies:

P = rainfall

I_v = canopy interception

Q_s = surface runoff

S_s = ground surface storage

Q_o – catchment outflow

E_o = open water evaporation

I_s = soil evaporation

Parameters for the unsaturated zone:

Q_{irr} = irrigation return flow

T_u = transpiration from the unsaturated zone

F = infiltration

S_u = storage in the unsaturated zone

Parameters for the saturated zone:

R = groundwater recharge,

Q_{lgi} = lateral groundwater inflow

C = capillary rise

S_g = storage in the saturated zone

Q_{gb} = groundwater seepage

Q_{lgo} = lateral groundwater outflow

Q_w = well abstractions

Q_{gf} = spring discharge

T_g = transpiration from the saturated zone

Fig. 3.1: The water balance components in a crystalline basement aquifer catchment.

A hypothetical catchment shall be used here to describe and explain the water balance of a small catchment in crystalline basement terrain. Fig. 3.1 summarizes the main features of such a catchment. In the figure Q denotes 'blue water[8]' fluxes, T denotes transpiration, (i.e., 'green water[9]'), S denotes storage, E denotes 'white water[10]' fluxes, g denotes the saturated (groundwater zone), u denotes the unsaturated zone, s denotes the ground surface zone and ow stands for open water as before.

---

[8] Water that is in the ground and surface water courses that can be exploited through physical interventions.

[9] This is the water used for biomass production by the green vegetation.

[10] This is the water that returns to the atmosphere from land and open water surfaces.

General assumption on the hypothetical catchment

Some assumptions can be made about the catchment of Fig. 3.1.

Human water withdrawals are minimal and can be neglected. Water abstractions are primarily for potable use and are small in comparison to natural fluxes in the hydrological balance, at least in the present socio-economic set-up.

The topography is undulating with a general decrease in slope from the topographical divide to the main stream channel such that both surface and subsurface flows are towards the main stream channel.

The catchment has one main stream channel into which surface water tributaries and groundwater feed such that the discharge from the catchment can be captured at a single catchment outlet.

No significant surface water impoundments, man made or natural, exists in the catchment such that surface storage is negligible compared to subsurface storages.

The aquifer is unconfined and recharged directly by rainfall. No other recharge sources exist.

Aquifer material is spatially heterogeneous but non-clayey such that soil suction and capillary rise are not dominant. Though layers can be identified in the aquifer, there is hydraulic continuity between them such that hydraulic behaviour at any point is aggregated over the different layers.

A crystalline basement underlies the aquifer such that deep vertical groundwater flow can be neglected.

The rainfall is seasonal and the stream discharge ephemeral.

These assumptions will guide the discussion of this discourse.

Hydrological system boundaries and partition points

The considered boundaries of the hydrological system are not very definite but they suffice for the purposes of evaluating a regional water balance. Four boundaries are considered. See Fig. 3.1.

The plane AD representing the earth surface defines the top boundary. The ground surface, open water body and any protruding surface representing any earth surface contact with the atmosphere form the top boundary of the catchment.

The plane BC representing the surface of the fresh crystalline basement rock that underlies the aquifer defines the bottom boundary. The top of this rock mass can be treated as an impermeable boundary.

The lateral boundaries, planes AB and CD, are defined by the topography, drainage and geology. The lateral boundaries present a dilemma as they are not only difficult to demarcate but also susceptible to 'seepage' losses of, or gains in, blue water especially for the saturated zone.

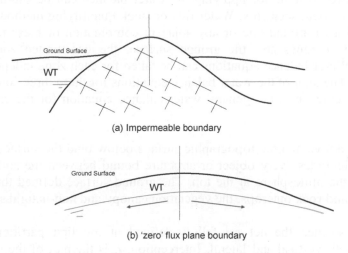

(a) Impermeable boundary

(b) 'zero' flux plane boundary

Fig. 3.2. Definitions of lateral boundaries at a topographical divide.

Fig. 3.2 shows the two possibilities for defining the lateral boundary of plane AB. Fig. 3.2 (a) defines the boundary in terms of geology. Since topographic high terrain is associated with parent rock material rather than the eroded deposits it can be assumed that an impermeable boundary exists at the topographical divide. Fig. 3.2 (b) defines the boundary from a groundwater flow perspective. Because the water table in an unconfined aquifer generally mimics the topographic surface a topographic divide is a zero flux plane for all flow perpendicular to the divide.

The two cases of Fig. 3.2 imply that the groundwater and surface water boundaries coincide with the topographical divide. Thus both surface and lateral sub-surface inflows at plane AB can be considered negligible.

The plane CD defines the interface between the groundwater flow and the surface flow system. When there is surface flow the surface water body defines a constant head boundary. If the head in the groundwater is above that in the surface water groundwater flows to the stream and vice versa. The plane CD does not represent a catchment boundary but a line of symmetry in that the processes highlighted in Fig. 3.1 are replicated on the right side of plane CD.

The water balance components are defined at imaginary partition points within the catchment. Three partitioning points can be defined. The first partition point is in the surface zone above plane AD and defines interception, infiltration and surface runoff. The second partition point is in the unsaturated zone below plane AD above the water table and defines transpiration and recharge (percolation). A third partitioning point occurs in the saturated zone above plane BC and defines capillary rise, fast groundwater discharge (spring flow) and slow groundwater discharge (baseflow).

The catchment water balance

A rainfall drop on the catchment will be progressively partitioned into fluxes and stocks as it moves from the earth surface into the aquifer and eventually out of the catchment through groundwater discharge. This partitioning can be assumed to occur at three hypothetical points in three distinct physical zones. See Fig. 3.1.

Any natural control volume for applying the water balance can be sub-divided into three congruent vertical segments. Water flux or stock quantifying methods are based on boundaries drawn around one or any adjacent combination of these three zones. The three physical zones are; the ground surface, the unsaturated zone and the saturated zone. Water balance equations can be taken for each zone independently of the other zones. The sum of the water balance equations for these three components at any given time scale gives the general water balance equation for the catchment at that time scale.

The atmosphere above and the topographic surface below bind the surface zone. The ground surface includes every object or structure bound between the soil surface at the bottom and the atmosphere at the top. The ground surface defined thus includes the bare ground and rock outcrops, the vegetation canopy and non-natural structures.

At the ground surface the net rainfall is defined at the first partitioning point. Partitioning is both vertical and lateral. Interception, I, is the part of the rainfall that accumulates on the surface and later returns to the atmosphere as vertical upward flux without penetrating the ground surface. In this case interception includes all fast evaporation processes including bare soil evaporation, evaporation from pools and evaporation from mulch (Savenije, 2004). For analysis purposes interception is assumed to occur from the soil surface, I_s, and vegetation canopy, I_v.

$$I = I_v + I_s \qquad\qquad (3.12)$$

Strictly speaking, interception is a storage but since it has a residence time usually less than a day it can be treated as a flux (Savenije, 2005).

The net rainfall is the difference between the total rainfall and the interception.

$$P_{net} = P - I \qquad\qquad (3.13)$$

The infiltration, F, is the portion of the rainfall that penetrates the ground surface as a vertical downward flux. All rainfall in excess of interception and infiltration, and is not stored on the ground surface, will constitute a lateral flux in the form of surface runoff, Q_s from the catchment.

The water balance for the surface zone is therefore given by:

$$\frac{dS_s}{dt} = P - \left(I_s + I_v + F + Q_s \right) \qquad\qquad (3.14)$$

where, S_s [L] and t [T] are the surface storage and time respectively. P [LT^{-1}] is the rainfall and all other fluxes have dimensions LT^{-1} and are as defined above.

The ground surface can be further sub divided into the soil surface and open water bodies. The open water bodies, permanent surface water bodies, ephemeral streams and temporary ponding systems can be considered separately as a subsystem of the ground surface. The important vertical fluxes in this case are the rainfall on the open water body, P_{ow} and the evaporation from the water body, E_{ow}. Lateral flows are the surface water contribution, Q_s, the groundwater contribution, Q_g and the catchment discharge, Q_o. The derived water balance can be written as:

$$\frac{dS_{ow}}{dt} = P_{ow} + Q_s + Q_g - (Q_o + E_{ow})$$ (3.15)

where, S_{ow} [L] is the storage in the open water body. Since the assumptions of this discourse are that the surface water system is ephemeral and of negligible areal extent in comparison to the unsaturated and saturated zones, the values of S_{ow}, P_{ow} and E_{ow} can be neglected and in a hydrological year for which storage change is negligible Equation 3.15 can be reduced to:

$$Q_o = Q_s + Q_g$$ (3.16)

where, Q_o is the total flow measured at the catchment outlet, Q_s is the 'overland' flow and Q_g is the groundwater flow. In situations where the surface flow is negligible catchment discharge is due to groundwater flow. Corollary, the catchment discharge gives an indication of the net recharge in the catchment.

The unsaturated zone proceeds from the ground surface and ends at the water table. The unsaturated zone is that portion of the hydrological system bounded by the ground surface (on top) and the water table (below). Water content in the unsaturated zone is thus below saturation. Inflows to the zone consist of downward fluxes from the surface zone; infiltration, F and irrigation excess, Q_{irr}, and the capillary rise, C, from the saturated zone below. The outflows from the unsaturated zone are predominantly vertical. Transpiration in biomass production, T_u is the major outflow whilst recharge to the saturated zone, R_u, is a secondary flux. The general water balance equation is given by:

$$\frac{dS_u}{dt} = (F + C + Q_{irr}) - (R_u + T_u)$$ (3.17)

where, S_u [L] is the storage in the unsaturated zone.

The ground surface and the unsaturated zone can be combined and analyzed as one medium. In such a case the water balance becomes:

$$\frac{dS_{s,u}}{dt} = P + C + Q_{irr} - (I_s + I_v + Q_s + R + T_u)$$ (3.18)

Note that if the capillary flow and irrigation flows are neglected and the evaporation fluxes are combined, Equation 3.18 reduces to the basic water balance equation for estimating groundwater recharge from the unsaturated zone.

$$R = P - Q_s - E - \frac{dS_{s,u}}{dt}$$ (3.19)

Thus recharge calculated from the unsaturated zone parameters can be defined as the difference between rainfall and the sum of the surface runoff, total evaporation and the change in surface and soil moisture storage.

The saturated zone is that part of the hydrological system which top boundary is the water table. The saturated or groundwater zone ends at the aquifer basement. For analysis purposes the thickness of the capillary zone between the unsaturated zone and the saturated zone will be ignored. The full water balance for the saturated zone can be defined by considering Equations 3.1 and 3.4.

$$\frac{dS_g}{dt} = \left(R + Q_{lgi}\right) - \left(Q_{gb} + T_g + Q_{gf} + C + Q_w + Q_{lgo}\right) \tag{3.20}$$

where Q_{gb} [LT^{-1}] and Q_{gf} [LT^{-1}] are the slow and fast groundwater flow components respectively. The lateral sub-surface flows can be assumed to be negligible from the system boundaries as defined above. In this case the balance equation reduces to:

$$\frac{dS_g}{dt} = R - \left(Q_{gb} + T_g + Q_{gf} + C + Q_w\right) \tag{3.21}$$

Equation 3.20 can be re-arranged and re-written as:

$$R = \frac{dS_g}{dt} + \left(Q_{gb} + T_g + Q_{gf} + C + Q_w\right) \tag{3.22}$$

Thus, groundwater recharge calculated from the saturated zone can be defined as the sum of the water outflows from the saturated zone plus the change in the saturated zone storage in any given time step.

Summation of equations 3.14, 3.15, 3.17 and 3.20 gives the full hydrological system balance equation as:

$$\frac{d}{dt}\left(S_{ow} + S_s + S_u + S_g\right) = P - \left(Q_{gf} + Q_w + Q_{gb}\right) - \left(T_u + T_g\right) - \left(I_v + I_s + E_{ow}\right) \tag{3.23}$$

where the total water stock in the catchment, S, is given by;

$$S = S_{ow} + S_s + S_u + S_g \tag{3.24}$$

the stream discharge from the catchment, Q_o, is given by;

$$Q_o = Q_{gf} + Q_w + Q_{gb} \tag{3.25}$$

the total transpiration (green water flux), T is given by;

$$T = T_u + T_g \tag{3.26}$$

the *direct* evaporation (white water flux), is given by;

$$I = I_v + E_{ow} + I_s \tag{3.27}$$

Equation 3.23 reduces to;

$$\frac{dS}{dt} = P - Q - T - I \tag{3.28}$$

The change in catchment storage is the balance of rainfall, stream discharge, catchment transpiration and catchment interception. Equation 3.28 shows that all hydrological processes in a catchment are driven by rainfall. A further simplification can be made by lumping the evaporative fluxes together.

$$E = T - I \qquad (3.29)$$

Equation 3.28 reduces to the classic catchment water balance equation:

$$\frac{dS}{dt} = P - Q - E \qquad (3.30)$$

Time scales of processes

The processes as described above do not occur at the same rate. There is therefore a need to have an idea of the time scales involved for each process. This is important when determining the storages and fluxes to include in a water balance at a selected time step. A time step greater than process scale implies that the process is accumulated in the time step and not a reflection of the actual process behaviour.

Table 3.1. Time scales of processes

Process	Flux (mm/d)		Relevant stock (mm)		*Time scale (days)	
	Symbol	Approx.	Symbol	Approx.	Formular	Approx.
Canopy Interception	I_v	3-5	S_s	2-3	$T_I = S_s/I$	0.5
Soil evaporation	I_s	1-2	S_s	1-5	$T_{Es} = S_s/E_s$	1-2
Open water evaporation	E_o	4-6	S_o	c1000	$T_{Eo} = S_o/E_o$	3-60
Surface runoff	Q_s	0.01	S_o	0.0+	$T_{Qs} = S_o/Q_s$	0.5-1
Infiltration**	F	50+	S_u	1000	$T_F = S_u/F$	<1
Transpiration from unsat. zone	T_u	1-2.5	S_u	1000	$T_{Tu} = S_u/T_u$	30-60
Recharge**	R	1-2	S_g	100	$T_R = S_g/R$	50-100
Transpiration from sat. zone	T_g	0.1-1	S_g	100	$T_{Tg} = S_g/T_g$	100+
Groundwater seepage	Q_g	0.2	S_g	100	$T_{Qg} = S_g/Q_g$	120

NB   *$T_F = S/F$, i.e. the time scale of a given flux, F, is its ratio to the relevant stock, S.
** Replenishing process

Some water balance dynamics

In a shallow groundwater system as is the case here, it is important to recognize that the magnitudes of the components of the water balance, particularly transpiration, are affected significantly by the position of the water table in relation to the root zone(s), which in turn is affected by the seasonality of the rainfall. In this regard three distinct cases can be considered.

In the dry season, the water table is below the root zone and only a minute number of trees namely the vlei type and riverine vegetation, the large indigenous fruit trees (e.g. Muhacha) and exotic trees (mainly, eucalyptus, pine and mango) tap from the groundwater resources. As most of the other indigenous trees shed their leaves in this period transpiration is correspondingly low. In the beginning (or towards the end) of the wet season the water table rises but an unsaturated zone exists and transpiration is from both the small and the large vegetation. At the peak of the wet season the water table reaches and, in some cases, overshoots the ground surface thereby completely eliminating the unsaturated zone for a considerable period during the wet season. In this case evaporation from the bare ground and surface water ponding systems dominate the vertical flux.

31

The catchment water balance therefore depends on the period of the hydrological year for which the balance is being considered. Some fluxes that are dominant during certain times are negligible during others.

3.2   The water balance approach and recharge estimation

All methods for quantifying hydrological parameters are based directly or indirectly on the water balance. In this section the methods applied in quantifying recharge will be discussed with respect to the water balance.

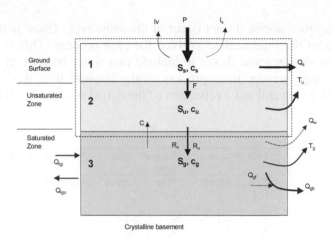

Fig.3.3. Schematic for applying the water balances to an arbitrary control volume.

Fig. 3.3 is a control volume of unit area taken from the catchment cross-section of Fig. 3.1. In the figure only dominant fluxes are considered. Spring flows and groundwater flows to river channels are combined into subsurface outflow, Q_g while irrigation return flows and groundwater abstractions are neglected. Transpiration from the saturated and unsaturated zones is represented by T_g and T_u, respectively.

3.2.1 Groundwater recharge from the water balance

The water balance technique for estimating recharge can be applied to the unsaturated zone and to the saturated zone.

The water balance for the combined unsaturated zone and ground surface can be written as:

$$\frac{\mathrm{d}S_{s,u}}{\mathrm{d}t} = P - \left(I_v + I_s\right) + C - T_u - Q_s - R \tag{3.31}$$

which can be rearranged to:

$$R = P - \left(I_v + I_s\right) + C - T_u - Q_s - \frac{\mathrm{d}S_{s,u}}{\mathrm{d}t} \tag{3.32}$$

Equation 3.32 is the expanded form of Equation 3.18. To be of practical use, the equation is simplified by assuming negligible surface runoff relative to the rainfall and the total evaporation and taking a long enough period such that the soil storage variation reduces to zero. In this case Equation 3.31 is reduced to:

$$R \cong P - E \tag{3.33}$$

Equation 3.33 is valuable as an indicator of the likelihood of recharge if E in the equation represents total evaporation. The equation estimates potential recharge and is most applicable in areas (or periods) in which rainfall is greater than potential evaporation.

If the water balance for the saturated zone is similarly considered, recharge can be estimated as:

$$R = Q_g + T_g + \frac{dS_g}{dt} \pm Q_{lg} \tag{3.34}$$

The subscript lg refers to the lateral subsurface flows into the control volume, Q_{lgi} and Q_{lgo}. Equation 3.34 can be simplified by assuming negligible direct evaporation from the groundwater reservoir and no lateral flows. Such situations occur if the catchment boundary coincides with a zero-flux boundary and if deep-rooted vegetation is mostly absent. The simplified form of Equation 3.34 then is:

$$R \cong Q_g + \frac{dS_g}{dt} \tag{3.35}$$

If Equation 3.35 is taken over several hydrological years a further assumption that the change in storage is small relative to the discharge can be made thus reducing the equation to the celebrated 'baseflow separation' technique.

$$R \cong Q_g \tag{3.36}$$

The assumption about storage changes limits the applicability of the 'base flow separation' technique to long-term series in which the influence of seasonal fluctuations is flattened out. Also, as can be discerned in Equation 3.36 the method is inaccurate in the case of catchments with 'evergreen' vegetation where T_g is non-negligible.

Equations 3.35 and 3.36 are of practical use at the catchment scale and offer a lumped and catchment aggregated recharge estimate. At a local scale a soil moisture balance in which only the water balance of the unsaturated zone is considered is often used. In such a case lateral flows are eliminated. The change in soil moisture is given by:

$$\frac{dS_u}{dt} = F + C - T_u - R \tag{3.37}$$

which, after substituting for F and re-arrangement, yields:

$$R = P + C - I - T_u - \frac{dS_u}{dt} \tag{3.38}$$

The recharge is estimated from the balance of the rainfall, P, capillary rise, C, the evaporative fluxes, I and T_u, and the change in soil moisture. This approach shows that the evaporative fluxes play an integral part in the estimation of the groundwater recharge. To avoid too many parameters in the determination of recharge simpler methods have been developed, the chloride mass balance technique, CMB, is such a method.

3.2.2 Groundwater recharge from the chloride mass balance (CMB) technique.

The basis of the method is the chloride mass balance between precipitation flux and the groundwater stock. As such, the method integrates the mass balance from the ground surface to the saturated zone per unit area. The water balance for the surface zone is given by:

$$\frac{dS}{dt} = P - \left(I_{veg} + I_s\right) - Q_s - F \tag{3.39}$$

Correspondingly, the mass balance with respect to chloride concentrations can be given by:

$$\frac{d\left(S_s \cdot c_s\right)}{dt} = P \cdot c_p - \left(I_{veg} + I_s\right) \cdot c_e - Q_s \cdot c_s - F \cdot c_s \tag{3.40}$$

Where c [ML^{-3}] denotes chloride concentration and the subscript e denotes evaporative fluxes while the upper case letters and subscripts are as defined in Fig. 3.1.

Equation 3.40 can be simplified if assumptions are made about the mass and water balances. First, the storage fluctuation term may be assumed negligible relative to inflows and outflows if the time of integration is taken long enough, in the order of several hydrological years, to cater for wet and dry years. As a result, the method only applies to long time scales and, thus, a multi-annual average. Secondly, water is assumed to evaporate in its pure form thus no chloride is lost through the evaporative fluxes. These two assumptions reduce Equation 3.40 to:

$$P \cdot c_p - Q_s \cdot c_s - F \cdot c_s \cong 0 \tag{3.41}$$

Equation 3.40 highlights the limitations of the chloride mass balance. Based as it is on the rainfall and groundwater chloride mass balance it implies that the surface runoff in Equation 3.40 has to be eliminated. It follows therefore that the chloride method can only be applied were the surface discharge is negligible such as in arid sand provinces where all the floodwater recharges the aquifer. But even if there is no surface runoff in a number of years, freak years in which Hortonian overland flow occurs during very intensive rainfall, can spoil the use of the method, since it will rinse the relatively saline water. Since the CMB method is only reliable over longer periods of time, there is always the risk that some exceptionally high rainfall occurs.

The rainfall intensity, diffuse sources, wind transport and the surface flow dynamics bring in other complications. High intensity rainfall usually generates runoff. The generated runoff can accumulate or dissipate chloride between the point of generation and the point of measurement. If the distance between these two points is large relative to the spatial scale of the recharge process distortions in the chloride concentration will arise nullifying the assumptions upon which the method is based. Because of this, the method is applicable in areas of high permeability and/or negligible topographical slope were most of the lateral flow ends up as recharge water within the balance boundary limits. If the horizontal flux is neglected Equation 3.41 is further reduced to:

$$P \cdot c_p - F \cdot c_s \cong 0 \tag{3.42}$$

Or simply,

$$F \cong P \cdot \frac{c_p}{c_s} \tag{3.43}$$

Equation 3.43 shows that rainfall is assumed to be the only source of chloride hence the chloride mass balance method is only applicable in areas where no alternative sources of chloride, such as salt deposit by wind or the use of fertilizers, exist.

Applying the chloride mass balance to the unsaturated zone yields the balance equation:

$$\frac{d(S_u \cdot c_u)}{dt} = F \cdot c_s - R \cdot c_u - T_u \cdot c_t \tag{3.44}$$

Equation 3.44 illustrates the difficulty associated in applying the chloride mass balance to the unsaturated zone alone. A way to overcome this difficulty is to combine the surface and the unsaturated zone by keeping the topmost boundary and describing a bottom boundary in the unsaturated zone not necessarily at the water table. For ease of analysis, the bottom boundary is made to coincide with the water table. In this case, F becomes an internal flux and is eliminated from the analysis.

This leaves two problems. Since the T_u in Equation 3.44 represents transpiration through plant soil moisture and nutrient uptake the process results in diminished chloride concentrations in the soil. This leads to an overestimate of recharge. Plants may exist that fix chloride ions in the soil resulting in an underestimate of recharge. The use of fertilizers further complicates the nutrient concentrations. Secondly, deep rooted vegetation may have their roots well below the chosen root zone depth, e.g., Eucalyptus trees, thereby distorting the value of T_u and the chloride concentrations in both the unsaturated and the saturated zones. For these reasons the chloride mass balance method is best applied were tree cover is either minimal or its nutrient processing dynamics are well understood. A favoured assumption however, has been that the effect of the soil mass and vegetation is minimal once steady state conditions have been achieved (Simmers, 1997). By measuring the downward flux below the root zone it is assumed that the effect of vegetation can be eliminated. This assumption also means that in the absence of lateral chloride inputs the chloride concentration just below the root zone is the same as that in the saturated zone thus c_u and c_g can be used interchangeably. In this case Equation 3.44 reduces to:

$$F \cdot c_s - R \cdot c_u = 0 \tag{3.45}$$

Or simply,

$$R = F \cdot \frac{c_s}{c_u} \tag{3.46}$$

If the change in chloride concentration in the control volume is negligible the mass balance reduces to the difference between Equations 3.43 and 3.46 which gives the standard chloride mass balance equation:

$$P \cdot c_p - R \cdot c_u = 0 \tag{3.47}$$

Or simply,

$$R = P \cdot \frac{c_p}{c_u} \tag{3.48}$$

The need for further restriction before applying the chloride mass balance equation is apparent from Equation 3.47. The groundwater chloride concentration in the saturated zone is assumed to be determined only by the vertical downward flux, in other words only from recharge. Thus the method is only applicable where lateral fluxes do not result in a change in chloride concentration. This is possible when there are no chloride sources in the saturated zone upstream of the control volume or there is no lateral flow. Alternatively, it should be possible to quantitatively account for the chloride input. Applying the chloride balance to the saturated zone explains this situation.

$$\frac{d(S_g \cdot c_g)}{dt} = R \cdot c_u + Q_{lgi} \cdot c_{gi} - T_g \cdot c_t - Q_g \cdot c_g - Q_{lgo} \cdot c_g \tag{3.49}$$

The vertical upward flux is removed from Equation 3.49 by assuming that direct evaporation from the groundwater is either negligible or for pure water only. The later assumption is dangerous in that direct evaporation from the saturated zone passes through the unsaturation zone and deposits some of its chloride in this zone thus c_t is non-zero in Equation 3.49. The chloride method is therefore of limited value in shallow groundwater systems where direct evaporation from the groundwater may be significant. The lateral flux can be eliminated as discussed above yielding:

$$\frac{d(S_g \cdot c_g)}{dt} = R \cdot c_u - Q_g \cdot c_g \tag{3.50}$$

Two scenarios then arise. Under steady conditions the change in storage is zero and the equation can be re-written as:

$$R \cdot c_u - Q_g \cdot c_g = 0 \tag{3.51}$$

As explained before, if c_u equals c_g then R equals Q_g.

Combining Equations 3.48 and 3.51 yields:

$$P \cdot c_p - R \cdot c_g = 0 \tag{3.52}$$

Or simply,

$$R = P \cdot \frac{c_p}{c_g} \tag{3.53}$$

Or,

$$Q_g = P \cdot \frac{c_p}{c_g} \tag{3.54}$$

Thus the chloride mass balance equation can be applied in the saturated zone to infer recharge were subsurface outflows can be measured, e.g., at springs only if the source of chloride is rainfall.

The second scenario occurs when outflows are assumed to be zero and the change in storage is therefore non zero.

$$\frac{d(S_g \cdot c_g)}{dt} = R \cdot c_u \tag{3.55}$$

This is the transient form of the chloride mass balance applicable were the vertical flux and the storage changes over a specific time period can be quantified. Such a scenario implies salt accumulation and shall not be discussed further in this discourse.

Considerations for dry deposition, capillary rise, interception and preferential flow

The chloride that falls to the ground surface in the absence of rainfall is defined as the dry deposition, D_s. When rain falls this dry deposition is added to the total chloride load and alters the chloride concentration of the infiltrating water. The steady state mass balance of the unsaturated zone under such conditions is given by:

$$R \cdot c_g = P \cdot c_p - (I_s + I_v) \cdot c_e - Q_s \cdot c_s - T_g \cdot c_t + D_s \tag{3.56}$$

where all parameters are as discussed above. When it is assumed that the concentration of chloride in evaporation fluxes is negligible and runoff can be neglected Equation 3.56 reduces to:

$$R = \frac{P \cdot c_p + D_s}{c_g} \tag{3.57}$$

Thus dry deposition increases the recharge estimate. Such an estimate is probably higher than the true recharge since the dry deposition can be attributed to sources other than the rainfall preceding the recharging event.

It should be noted that the analysis explained so far applies only to predominantly recharge areas, i.e., in areas where flow is vertically downwards. In areas where capillary rise is significant, vertical flow is bi-directional, i.e., both downwards from percolating rainfall and upwards from the saturated zone. In this case the net downward flow, net recharge, has to be considered. Equation 3.56 is modified to:

$$(R - C) \cdot c_g = P \cdot c_p + (I_s + I_v) \cdot c_e - Q_s \cdot c_s - T_g \cdot c_t + D_s \tag{3.58}$$

where, C [LT^{-1}] is the capillary rise.

If the same assumptions on evaporation and surface runoff above are retained Equation 3.58 yields:

$$(R - C) = \frac{P \cdot c_p + D_s}{c_g} \tag{3.59}$$

Equation 3.58 is the same as Equation 3.56. The impact of capillary rise is to increase the chloride load and if it is not considered erroneous recharge estimates are obtained.

A recent trend in the application of the CMB method is to consider interception and preferential flow in the above equations. The underlying arguments are that interception reduces the water volume whilst preferential flow recharge takes the rainfall chloride concentration directly to the saturated zone. However, a simple derivation can show that such considerations are incorrect.

Interception changes the concentration of the net rainfall. Interception, as explained before is a "pure" flux. All the chloride initially in the rainfall will be in net rainfall. Thus by the conservation of mass:

$$P \cdot c_p = P_{net} \cdot c_s \tag{3.60}$$

which can be reduced to:

$$c_s = \frac{P}{P - I} \cdot c_p \tag{3.61}$$

The chloride concentration in net rainfall is thus higher than that in gross rainfall.

Groundwater recharge is due to net rainfall. The recharge estimate when interception, I is considered is given by:

$$R = \frac{P_{net} \cdot c_s}{c_g} \tag{3.62}$$

which, after substitution of Equation 3.61 reduces to:

$$R = \frac{P - I}{c_g} \cdot \frac{P}{P - I} \cdot c_p \tag{3.63}$$

and simplifies to:

$$R = \frac{P}{c_g} \cdot c_p \tag{3.64}$$

which is the same equation as Equation 3.53.

Thus interception does not influence the recharge estimate since the reduction in the rainfall is compensated by the increase in the chloride concentration in the net rainfall.

The impact of preferential flow can be similarly considered. Fig. 3.4 shows the general case of Fig. 3.3 when preferential flow, R_s, is incorporated.

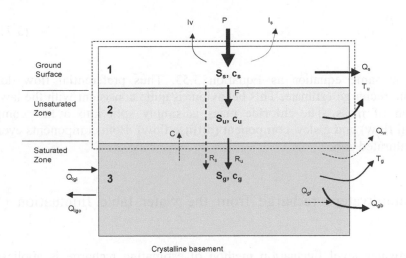

Fig. 3.4. Schematic for applying the water balances to an arbitrary control volume with consideration for preferential flow.

The chloride mass balance for the saturated zone in the case of Fig. 3.4 is given by:

$$R \cdot c_g = R_u \cdot c_u + R_s \cdot c_s \tag{3.65}$$

where R is the total recharge, R_u is the diffuse recharge and R_s is the recharge due to preferential flow. Usually an underlying assumption that the chloride concentration in the preferential flow is the same as that in the rainfall is made. This assumption is not correct since preferential flow has the chloride concentration of net rainfall rather than that of gross rainfall. As such Equation 3.60 above gives the chloride concentration in the recharge water.

The diffuse flow can be taken as a fraction, p, of net rainfall. In this case the chloride mass balance for the diffuse flow component will be given by:

$$R_u \cdot c_u = p(P - I) \cdot c_s \tag{3.66}$$

which, after substituting for c_s reduces to:

$$R_u \cdot c_u = p \cdot P \cdot c_p \tag{3.67}$$

Similarly the chloride mass balance for the preferential flow component is given by:

$$R_s \cdot c_s = (1 - p) \cdot (P - I) \cdot c_s \tag{3.68}$$

which, after substituting for c_s reduces to:

$$R_s \cdot c_s = (1 - p) \cdot P \cdot c_p \tag{3.69}$$

Substituting Equations 3.68 and 3.69 into Equation 3.66 gives:

$$R \cdot c_g = p \cdot P \cdot c_p + (1 - p) \cdot P \cdot c_p \tag{3.70}$$

which, after rearrangement reduces to:

$$R = \frac{P}{c_g} \cdot c_p \tag{3.71}$$

which is the same equation as Equation 3.53. Thus preferential flow does not influence the recharge estimate. This behaviour is quite consistent with the law of the conservation of mass. The chloride mass is simply split into a fast component (preferential flow) and a slow component (diffuse flow). Both components eventually reach the saturated zone.

3.2.3 Groundwater recharge from the water table fluctuation (WTF) technique.

The groundwater level fluctuation method of estimating recharge is applicable by definition to the saturated zone. A catchment water balance for the saturated zone can be written as before:

$$\frac{dS_g}{dt} = R + Q_{lgi} - Q_{lgo} - T_g - Q_g - Q_{sp} + C + Q_w \tag{3.72}$$

Assumptions about spring discharges, well abstractions and capillary rise can be restated reducing Equation 3.72 to:

$$\frac{dS_g}{dt} = R + Q_{lgi} - Q_{lgo} - T_g - Q_g \tag{3.73}$$

To reduce the number of terms in Equation 3.73 further, more assumptions have to be made. The first assumption is that direct evaporation from the groundwater is insignificant which fact does not hold in shallow groundwater systems in arid climates were the potential evaporation is high. The second assumption is that upward vertical fluxes from deep underground aquifers can be ignored. This is valid where the aquifer has an impermeable base as in crystalline basement aquifers or in confined systems. Accepting these simplifying assumptions leads to Equation 3.74.

$$\frac{dS_g}{dt} \cong R + Q_{lgi} - Q_{lgo} - Q_g \tag{3.74}$$

Horizontal flow in the saturated zone is governed by Darcy's law.

$$Q_g = k \cdot \frac{d\varphi}{ds} \tag{3.75}$$

Where Q_g is the horizontal groundwater flow, k [LT^{-1}] is the hydraulic constant and $d\varphi$ [L] is the change in hydraulic head and ds [L] is the traversed distance along a flow path. Assuming aquifer homogeneity, Equation 3.76 can be written in terms of Darcy as follows:

$$\frac{dS_g}{dt} \cong R + k \cdot \left(\left(\frac{d\varphi}{ds} \right)_{lgi} - \left(\frac{d\varphi}{ds} \right)_{lgo} \right) - Q_g \qquad (3.76)$$

If the hydraulic gradient does not change on the side boundaries of the control volume then the horizontal inflows equal the horizontal outflows and Equation 3.76 reduces to:

$$\frac{dS_g}{dt} \cong R - Q_g \qquad (3.77)$$

The recharge can therefore be estimated from the changes in storage in the groundwater zone. Equation 3.77 can be rearranged to:

$$R \cong Q_g + \frac{dS_g}{dt} \qquad (3.78)$$

Equation 3.78 is applicable if the groundwater outflow is either a point discharge such as a well-defined spring or the interchange between the groundwater system and the river channel can be accurately defined. The discharge and groundwater levels can be used to calculate recharge provided the storage characteristics, e.g., specific yield of the aquifer are well understood. Equation 3.78 is a special case of the 'baseflow separation' technique in which the storage changes in the discharging aquifer are accounted for.

The method has to be applied cautiously since groundwater fluctuations are not always entirely due to recharge. Factors like abstraction, atmospheric pressure changes and transipiration can induce groundwater level changes, (Todd, 1980). The method is therefore most suited where groundwater levels are shallow, easily discernable and due only to recharge events.

3.3   Summary and conclusions

It has been shown that recharge is an integral part of the water balance and all techniques for estimating recharge can be explained on the basis of the water balance. It has been shown that common errors in the water balance techniques can be explained and eliminated if the water balance is applied correctly. However, simplifying assumptions are required in applying the water balance. If these are misunderstood erroneous conclusions can be reached about the hydrological processes. It should be noted however, that even when the water balance is applied correctly and the right assumptions made measurement errors still remain the main problem in estimating recharge from water balance components.

Table 3.2 below summarizes the main formulae, application scales and basic assumptions behind the recharge estimation methods based on a water balance analysis.

Table 3.2. Recharge estimation methods from water balance perspective.

METHOD	FORMULAR FOR RECHARGE (R)		SPATIAL SCALE	CONDITIONS & ASSUMPTIONS
Direct water Balance	Unsaturated zone	$R = P - Q_s - E - \dfrac{\mathrm{d}S_{s,u}}{\mathrm{d}t}$	Catchment + Local	
	Saturated zone	$R = \dfrac{\mathrm{d}S_g}{\mathrm{d}t} + \left(Q_{gb} + T_g + Q_{gf} + C + Q_w\right)$	Catchment + Local	Lateral flows negligible
Chloride deposition	Unsaturated zone	$R = P \cdot \dfrac{c_p}{c_u}$	Point	Evaporative fluxes have zero concentration. Rainfall is the only source of chloride.
	Saturated zone	$R = P \cdot \dfrac{c_p}{c_g}$	Point + (Local)	As for unsaturated zone + No lateral flow.
Groundwater level fluctuations	Recharging: short period	$R = \dfrac{\mathrm{d}S_g}{\mathrm{d}t}$	Local + (Point)	Fluctuations due solely to recharge. Aquifer parameters known.
	Recharging: Long period	$R = Q_g + \dfrac{\mathrm{d}S_g}{\mathrm{d}t}$	Catchment + local	Fluctuations due solely to recharge. Discharge flows measurable. Aquifer parameters known.

CHAPTER FOUR

4.   BACKGROUND TO THE STUDY AREA

In this chapter a summary of the general background of the country and of the Mupfure river basin is given. Fig. 4.1. shows the political and drainage map of Zimbabwe and the location of the Mupfure and Nyundo catchments.

4.1   General background to Zimbabwe

Zimbabwe is a landlocked country in Southern Africa, located between the latitudes 15° 30′ and 22° 31′ South of the Equator and between the longitudes 25° 30′ and 33° 10′ East. The country lies wholly within the tropics.

Fig.4.1. Map of Zimbabwe and the location of the Nyundo catchment (rectangle).
(Source:)

The climate in Zimbabwe can be described as moderate. Two seasons can be defined, a wet and hot season from October to March and a dry and cold season from April to September. Maximum day temperature in the warm season rises to around 30°C in the high veld and 35-40°C in the low veld. During the cold period most areas in Zimbabwe may experience frost and the average temperature is around 15°C.

The seasonality influences the water resources, vegetation and socio-economic characteristics in the country. For example, almost all rivers and streams are ephemeral, most vegetation is deciduous and sheds leaves in the dry season whilst agriculture is rain fed and takes place in the wet season.

Geology, soils and vegetation

The geology of Zimbabwe is dominated by the Precambrian Basement Complex composed mainly of granite-greenstone belts over 3000 million years of age (Stagman, 1978). The Basement Complex covers over 70% of the country and is associated more with mineral deposits than groundwater resources. For example, The Great Dyke running across central Zimbabwe is host to one of the world's largest known chromium resources. The groundwater potential is limited unless there is substantial weathering and/or fracturing.

The next significant geological formation consists of the Karoo Basins which cover 15% of the country. The Karoo formations originate from lava and sediments from volcanic eruptions and erosional activities. Groundwater potential is poor unless there is substantial weathering.

The Kalahari System covers 12% of the country and originates from aeolian deposits. The depth of the unconsolidated material in these formations gives rise to substantial groundwater potential.

Natural soils in Zimbabwe are predominantly derived from granite of the Basement Complex and are often sandy, light-textured and of only fair agricultural potential. Having relatively low inherent fertility, increased production can only be achieved through good land management and the application of fertilizers and animal manure. Such farming practices combined with the unconsolidated nature of sandy soils imply a greater risk of groundwater contamination through leaching.

Zimbabwe is characterized by Savanna[11] woodland interspersed with open grassed drainage lines or dambos[12]. Except in the dambos, the climax vegetation throughout Zimbabwe would be forest but the natural vegetation has been much modified by fire and by the replacement of indigenous woodland by crops, pastures and exotic trees. Less than a tenth of the land area remains under closed canopy forest or plantations while about half retains patchy tree cover on grazing land. The vegetation variety is primarily determined by differences of altitude and rainfall but may be overridden by edaphic factors, mainly soil fertility. Miombo and Mopane are the most extensive woodlands (Msipa, 1992).

---

[11] Grasslands with trees of moderate size.
[12] Localized wetland, often a discharge area in CBAs.

44

Hydrology and water resources

The largest portion of annual rainfall in Zimbabwe is convectional. The rain normally appears as heavy showers frequently accompanied by thunderstorms resulting in high intensities. Mean annual rainfall for the whole country is about 675 mm/a. Practically, all of the annual rainfall falls in the wet/hot period. The coefficients of variation of annual rainfall lie between 20 and 40 percent, with the highest values prevailing in the low rainfall areas to the southwest of the country (MEWRD, 1985).

Evaporation is high in Zimbabwe. Annual open water evaporation ranges between 1,400 and 2,200 mm/a. The spatial and inter-annual variations in evaporation over the country are lower than for rainfall. The annual open water evaporation is higher than the mean annual rainfall, and most of the rainfall evaporates.

On average only approximately 8 percent of the rainfall finds its way to the rivers as runoff. (MEWRD, 1985). The total runoff amounts to approximately 20 000 million cubic metres per year, or 52 mm/a. Mean annual runoff however, varies greatly over the areas of Zimbabwe, the figures ranging from less than 15.7 mm/a in the low rainfall areas to more than 157.8 mm/a in the high rainfall areas (MEWRD, 1985). In some areas up to 70% of the total runoff can be attributed to baseflow (Savenije, 1998; Butterworth et al., 1999). By deduction and assuming no net annual groundwater storage change, average annual groundwater recharge can be estimated at 5% of annual rainfall.

Though the average annual rainfall in Zimbabwe is above the world average and the per capita availability above normal, the country remains water stressed for several reasons. The low runoff coefficient, less than 10%, results in low volumes of blue water. The seasonal nature and high inter-annual variations demand large and expensive storage facilities to avail the resources in periods of scarcity. High evaporation losses, on the other hand, means only 50% of the mean annual runoff is available from surface impoundments (MLWR, 1984).

The country does not possess good aquifers as much of it is covered by the Basement Complex as described above. Consequently, groundwater availability is severely constrained. However, because of its in-situ nature groundwater remains the most dependable and in some cases the only source of water particularly for rural communities (MLWR, 1984; Mudzingwa, 1998). The country's groundwater resources are low to moderate. Exploitable quantities are highly variable in space and time. In much of the basement complex areas aquifers tend to be shallow, localized and dependent on annual rainfall for replenishment such that drought years are generally associated with dry wells (Butterworth et al., 1999). In the Kalahari sand areas, aquifers are deeper, have low variability but expensive deep wells are required to access the water.

By 1990, the per capita water availability for the country stood at 1776 m^3 (Chenje & Johnson, 1996). This figure is expected to drop to 1150 m^3 by the year 2025 (Hirji et al., 2002) making the country water stressed. The biggest water user is agriculture which accounts for 79 % of all water uses followed by domestic consumption at 14 % and industry at 7 % (Hirji et al., 2002). However, such a distribution is misleading in that the importance of water is reflected more by the population depended on the water than the volume of water used. In this sense primary water use is the most important of all water uses. Conversely, groundwater is the most important of all water sources in the country meeting the basic water requirements for up to 70% of the national population (Martinelli & Hubert, 1985).

45

Water resources management

Water law and management in the country have evolved over a century from a traditional non-formal system through a colonial appropriation approach to a more reformist and egalitarian approach of the post-colonial era.

The prior-appropriation doctrine was applied for over ninety years and culminated in the 1976 Water Act. The main features of the act were a prior date system of allocation administered by the administration court, treatment of groundwater as private water and a water board system of water management (Murungweni, 2001).

A new Water Act of 2000 repealed the 1976 Water Act, established a self-financing national water authority, set up water user councils based on hydrological catchment boundaries, made all water public and set priorities for water use with primary and environmental use having the highest priority.

In the communal areas where water is, in the majority of cases, exclusively for portable use water management is based on stakeholder participation in the form of non-formal water point associations. Such organizations often lack the full hydrological knowledge to enable them to fully exploit their resources especially in times of hydrological scarcity.

4.2   The Nyundo catchment

The Nyundo catchment covers approximately 180 square kilometres in the Mhondoro communal areas of central Zimbabwe. The catchment is situated approximately 100 kilometres south of Harare the capital city of Zimbabwe. The catchment forms part of the headwater catchment for the eastern part of the Mupfure catchment and contributes 1.5 % of the mean annual runoff in the Mupfure catchment.

The Mupfure catchment covers an area of 12,050 square kilometres between longitudes 290° 40′ and 31° 30′ East and latitudes 17° 50′ and 18° 55′ South. The general elevation ranges from 1500 metres in the upper part of the catchment to about 750 metres at the confluence with the Munyati River. The Mupfure catchment covers 1.5 % of the land area of the Zambezi River Basin and contributes about 1.2 % of the total runoff in the larger basin.

The Zambezi River Basin drains six countries and is the second largest river basin in the southern Africa sub-region. The basin covers an area of 1 942 700 Mm^2 and discharges an average of 94 000 Mm^3/a into the Indian Ocean.

The topography is almost flat to gently undulating with the majority of slopes less than 2 percent. The elevation over most of the area is 1 250 to 1 350 metres above mean sea level.

A demographical survey carried in late 2001 showed that the catchment is home to about 5 000 people with an average household size of about 7 people. Fig. 4.2 summarizes the population characteristics in the catchment.

The major crops cultivated in the area are maize (70% of total) followed by groundnuts (21%) and wheat (8%). About 50% of all produce is for internal consumption whilst the remainder was sold.

Water use is mostly primary[13] with average per capita water consumption around 25 lcd[14]. The most dominant water source is the shallow well. Most water points are within 500 m of the point of use. Total catchment consumption is around 155 m³/d.

A more elaborate description of the water resources and hydrogeology of the Nyundo catchment is given in Chapter 5.

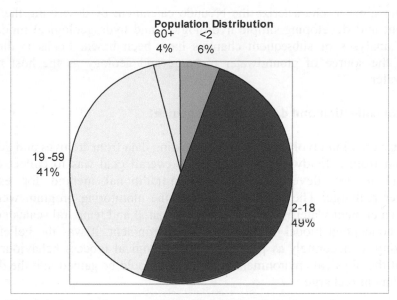

Fig. 4.2. Population age distribution in the Nyundo catchment.

---

[13] Water used for drinking, cooking and washing.
[14] Litre per capita per day

CHAPTER FIVE

5. FIELDWORK AND DATA INTERPRETATION

This chapter aims to characterize the Nyundo catchment by describing the physical environment and developing simple hydrological and hydrogeological models upon which the analyses of subsequent chapters have been based. Focus is directed at rainfall as the source of groundwater recharge and geology as the host media to recharge water.

5.1 Data collection and database management

The database creation involved collation of existing data from archives and generation of new data from a fieldwork campaign. The overall goal was to collect data that would lead to the development of a non-traditional method for estimating groundwater recharge. The main objectives of the monitoring program were (1) to define the catchment water balance at different spatial and temporal scales and (2) to develop a conceptual model of the physical environment. It was the belief that by characterizing as accurately as possible the hydrological process behaviour and the response of the physical environment to it, insight would be gained into the dynamics of the catchment recharge.

5.1.1 The monitoring network

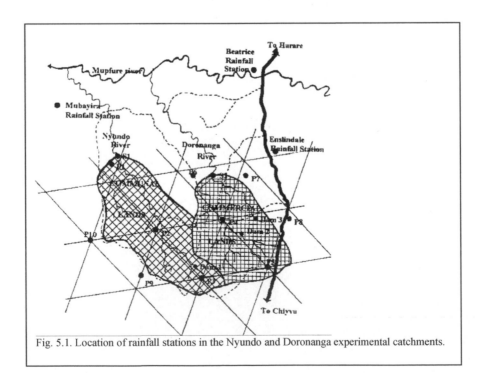

Fig. 5.1. Location of rainfall stations in the Nyundo and Doronanga experimental catchments.

Fig. 5.1 shows the rainfall monitoring stations. The final network consisted of 9 rain gauge stations located at a distance of approximately 11.5 km from each other in a triangular network as recommended by Askew (1995). The network covered approximately 940 Mm^2 giving a gauge density of 1 gauge per 104 Mm^2 or 9.6 gauges per every 1 000 Mm^2. Each station consisted of two rain gauges, an automatic recorder and a standard PVC collector rain gauge of 100 mm capacity.

Apart from rainfall, stream discharge, groundwater levels and open water evaporation were also monitored. Fig. 5.2 shows the monitoring network. The automatic recorders took water level and temperature readings at an hourly interval. Manual records of groundwater levels were collected at a bi-weekly interval.

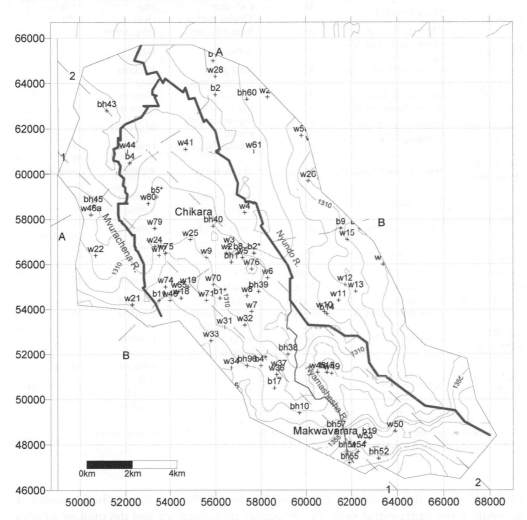

Fig. 5.2. Base map for the Nyundo catchment showing the monitoring sites, topography, drainage and hydrogeological sections.

5.1.2 Data collection

Data collection took four years from 1998/99 to 2002/03. Data sources included national hydrological databases, fieldwork and literature references where raw data could not be obtained. Three groups of data were considered, study period continuously monitored data (automatically recorded), study period manually collected data and long period regional data, in most cases manually recorded.

Table 5.1. The data collection and monitoring campaign

Process/state	Parameter	Gauging Method	Instrument	Frequency	No of sites
Rainfall	Precipitation, P	M	Collector	Day	2 (9)
		A	Tipping bucket	Hour	2 (9)
	Chloride content, c_p	M	Rain collector	Rain events	1
Evaporation	Open water evaporation, E_o	M	A-pan	Day	1
Discharge	Catchnment discharge, Q_o	M	Staff gauge	Random	1
		A	Pressure sensor	Hour	1
	Dry season flow, Q_g	M	V-notch	Random	3
	Spring discharge, Q_{sp}	M	Jug + clock	Random	2
Groundwater storage	Groundwater level fluctuations, dh	M	Dip meter (light + sound)	Bi-weekly	30
		A	Pressure sensor	Hour	2 (4)*
Groundwater abstractions	Human abstractions, Q_w	M	15 litre bucket	Random	60 questionnaires
Water chemistry	Miscellaneous. ions	---	"grab" bottles	Random	20
		M	Paqua lab.	2 times	20
	Groundwater chloride	M	"grab" bottles	Random	20 (repeated)
Topographic surface	Levels, m.a.s.l	M (C)	Tacheometry	1/study period	21
Vegetation cover	Spatial extent	M (C)	GPS	Random	15
	Tree densities	M (C)	Counts		
Soil properties	Porosity, n	--- (C)	Auger	Random	12
	Grain distribution, ϕ	--- (C)	"Grab" samples	Random	30
Aquifer properties	Permeability, K	M (C)	Slug test (dip meter, timer, bucket)	Random	10
Geological distribution	Spatial extent	M (C)	GPS	Random	various
			VES	3/study period	15 + 10 + 20
			Drilling	Random	8

NOTES: M = Manually collected data A = Automatically collected data
C = Complementary data   * = includes sites with short or suspect records.
VES = vertical electrical sounding   GPS = Global positioning system.

Table 5.1 shows the parameters that were monitored, the gauging method (manual or automatic), the instruments used, the frequency of monitoring and the number of sites monitored for the different parameters. The process/state column represents the hydrological property or physical phenomenon being considered whilst the parameter column represents the monitored variable. The gauge method column illustrates whether the data was collected directly by human monitors (M) or was recorded continuously (A) using data loggers. Complementary data (C) is data not requiring sustained monitoring. The frequency column indicates how often the data was collected. The recording times were synchronized for all gauges to enable correlation of readings.

Long period records were collected from the databases of government departments mainly from the Hydrology branch of the Zimbabwe National Water Authority (ZINWA). The Beatrice record stretches for 43 years from 1952/53 to 1994/95 whilst the Mubayira record covers 31 years from 1969/70 to 2000/01. The records only give monthly data and annual totals. Rainfall data included the national long-term monthly averages for all stations in the country from 1961 to 1991 (thirty years).

Long period flow data was obtained for stations C70, C67 and C66 on the Mupfure River of which the Nyundo catchment is a sub-basin. Records cover the 1960/61 to 1995/96 period.

Measuring equipment

Automatic rainfall measurements recorded using a data logger[15] that uses a tipping bucket mechanism and records to an accuracy of 0.1 mm. 100 mm capacity PVC collector rainfall gauges were used for manual rainfall measurements.

Automatic measurements for groundwater level fluctuations and temperature were recorded using pressure transducers[16] that measures water and atmospheric pressure in centimeters water column with a range of zero to 500 cm. The data is recorded in the logger at specified time intervals and can be downloaded to a computer at a future date. By recording both the water and atmospheric pressure the water depth could be estimated for pooled surface water and for groundwater. Each diver station was equipped with two divers one immersed in water to record the water pressure and the other suspended above the water table to measure the atmospheric pressure. Manual groundwater levels were measured using a dip meter attached to an electrical sensor device that makes a sound and lights a bulb on contact with water (Hudak, 2000).

River discharge was measured at the catchment outlet gauging weir using two divers positioned in such a manner that the bottom diver sensor coincided with the zero reading of the flow gauge plate.

5.1.3 Data processing

Hourly rainfall was algebraically summed to give daily totals.

$$P_{day} = \sum_{i=1}^{i=24} (P_{hr})_i \qquad\qquad (5.1)$$

Where, P_{day} (mm/d) is the daily rainfall, P_{hr} (mm/hr) is the hourly rainfall and i is an hour counter. Similarly the totals for monthly and annual rainfall were calculated.

Pressure data must be converted to water level data to obtain groundwater levels by subtracting the atmospheric pressure reading from the corresponding water pressure reading.

---

[15] A Rainman logger by Van Essen Instruments was used.
[16] MTD divers also developed by Van Essen Instruments were used.

$$h_p = \left(\frac{P_w - P_a}{100} \right) \qquad (5.2)$$

where h_p is the height of the water column (m), P_w and P_a are the water and atmospheric pressures (cm) respectively and 100 (cm/m) is a conversion factor from cm to m.

As hp is a state variable rather than a flux variable, the water level for a day was taken as the level at a specified time rather than an average for the day. Consequently all daily groundwater levels refer to the level at 08h00 on a given day. Similarly the monthly level is the level on the 30th day of each month except for February for which the reading is for the 28th day.

The discharge estimation required a slightly different approach to the treatment of the water levels.

$$h = k \cdot \left(\frac{P_w - P_a}{100} \right) \qquad (5.3)$$

Where h (m) is the water stage as read on the gauge plate, P_w and P_a are as defined before and k [-] is a proportionality constant between the water height estimated from pressure readings and the manual gauge plate reading. A constant of 0.45 has been used in this study. The constant accounts for pressure increases due to the kinetic energy of the flowing water. In groundwater the flow velocity is low and the influence of kinetic energy on water head is negligible.

The discharge is read from a rating curve specifically developed for the Nyundo weir.

$$Q = f(h) \qquad (5.4)$$

Where Q is the discharge in m³/s and h is the stage in m.

The accumulation of the data to daily discharge was done in a two stage process to obtain hourly discharge first then summing the hourly data over a 24 hr period to obtain the discharge at 08h00 for each day.

$$Q_{hr} = 3600 \cdot Q_{sec} \qquad (5.5)$$

and,

$$Q_{day} = \sum_{i=1}^{i=24} (Q_{hr})_i \qquad (5.6)$$

Temperature values were taken directly from the diver records.

Data filling and correlation

Correction and filling of missing data was accomplished through two simple techniques.

For process variables, rainfall and discharge double mass analysis (Dahmen & Hall, 1990) was utilized to estimate missing values at given stations at the same time if a linear fit was observed with the cumulative mass plots. The following regression equations were then used to estimate missing or suspect readings.

$$Y_n = b \cdot X_n \tag{5.7}$$

and,

$$Y_n = \sum_{i=1}^{i=n} y_i ; X_n = \sum_{i=1}^{i=n} x_i \tag{5.8}$$

and,

$$x_n = X_n - X_{(n-1)} \tag{5.9}$$

Where upper case letters refer to the cumulative sum of the readings at a point and the smaller case letters refer to the recorded value at the point and i and n denote the length of record.

State variables were estimated by averaging in time, Equation 5.10, and interpolating in space, Equation 5.11, assuming a linear trend.

$$h_i = h_o + \left(\frac{h_n - h_o}{n} \right) \cdot i \tag{5.10}$$

where, h (m) denotes water levels, 0, and n denote the start and end of the missing record and i is the number of required value within the missing record.

$$h_i = h_{(i-1)} + \left(\frac{h_{i+1} - h_{i-1}}{l_{i+1} + l_{i-1}} \right) \cdot l_{i-1} \tag{5.11}$$

where h (m) denotes water level, l (m) denotes the spacing between stations and the subscripts denote forward and backward neighbourhood stations with known water levels.

5.1.4 Screening checks and data reliability

As data was for a relatively short hydrological time series screening checks were limited to tabular comparisons, mass curve analysis, inspection of time series plots and spatial correlation checks (spatial homogeneity). Only long period rainfall and discharge data was subjected to trend and stability analysis.

Fig. 5.3 shows that the automatic data correlates well to the manual data with coefficients of correlation between 85% and 98%. Mass curve comparison shows that the data is reasonably consistent for a hydrological analysis.

A pair wise comparison of data from different stations showed correlation coefficients between 75% and 95%.

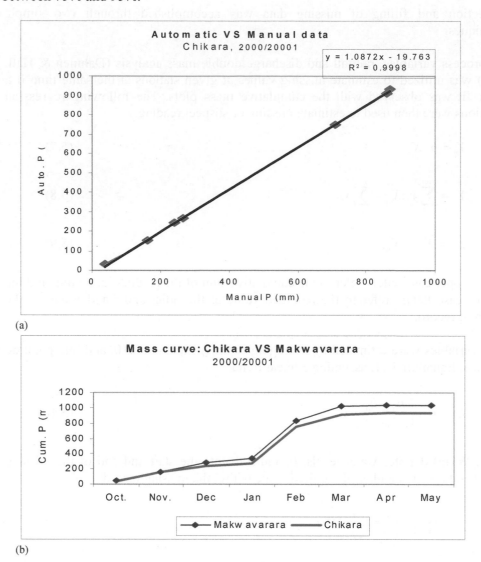

(a)

(b)

Fig 5.3. Data reliability plots. (a) Correlation of manual and automatic data for Chikara. (b) Mass curves for Chikara and Makwavarara in 2000/2001.

5.2   Rainfall characteristics

In this section the rainfall and stream discharge are analysed. The main objectives are (1) to describe the hydrological behaviour of the catchment and its surroundings and (2) interpret the hydrological observation in relation to recharge occurrence and estimation.

The spatial and temporal availability of the rainfall and stream discharge are discussed at annual to daily time steps. Probability and frequency distributions are suggested. The analysis is based on both study period data from the Nyundo catchment and long period data from the larger Upper Mupfure catchment.

Annual and monthly time steps are larger than the time scales of most hydrological processes. As such the actual behaviour of those processes is masked or completely lost in analysis at monthly or annual time steps. The daily time step is closer to the time scale of most hydrological processes such that analyses at this time step can offer more realistic insights into the behaviour of those hydrological processes.

A rainfall event in the region has a time scale in the order of a few minutes to several days depending on the type of rainfall. Direct recharge is anticipated to have similar time scales or slightly larger depending on the recharge mechanism. Preferential flow recharge may have the same time scale as the recharging rainfall event whilst diffuse flow recharge may exhibit a time lag between rainfall occurrence and groundwater system response. Thus comparing the behaviour of rainfall and recharge at a daily time step answers two important questions in recharge estimation. (1) When does recharge occur in relation to rainfall and, corollary to this question, (2) by what mechanism does the recharge occur?

A quick response in the groundwater system to rainfall occurrence implies preferential flow mechanisms are dominant and that direct rainfall-recharge relationships in similar manner to runoff may be feasible.

Important parameters of daily rainfall that influence recharge are rainfall depth, duration, and intensity of a given rainfall event and the temporal and spatial distribution of the rainfall. Current thinking, especially in water resources modeling, is that previously ignored rainfall descriptors like number of rain days and spell lengths are equally important in describing the behaviour of daily rainfall and may improve greatly the predicative functions of hydrological models at larger time steps (De Groen, 2002).

The ensuing discussion attempts to explain the daily rainfall characteristics in the Nyundo catchment in as far as they may influence recharge behaviour and therefore recharge estimation techniques.

5.2.1 Rainfall formation

The rainfall in the SADC region is both convectional and cyclonic (Torrance, 1981; Makarau, 1995; De Groen, 2002). The inter-Tropical Convergence Zone (ITCZ) and the Indian Ocean Cyclones (IOC) have been accepted as the two main causes of rainfall in the region.

The ITCZ is the zone where air streams originating in two hemispheres, the northwest and southeast, meet. The southern edge of the sub-tropical high pressure moves south by 3 to 4 degrees latitude whilst the Indian Ocean high pressure moves east by about 25 degrees longitude (Gieske, 1992). The convergence within the ITCZ leads to a decrease in stability over the southern sub-continent which results in convectional rainfall as warm air is pushed upwards. The phenomenon results in thunderstorms and showers of limited daily spatial extent, usually between 5 and 10 Mm^2 (De Groen, 2002). Up to 90% of the rainfall occurring in Zimbabwe between October and January is related to the ITCZ (Torrance, 1981).

The IOC brings rainfall mostly to coastal areas of the sub-region but sometimes causes rainfall in the hinterland. Such cyclones account for part of the rainfall between December and April (Makarau, 1995).

The pattern of rainfall in the region has implications on the occurrence of groundwater recharge. Seasonal rainfall implies recharge occurs at specific periods in the hydrological year whilst the highly variable nature of the rainfall is likely to be replicated by the recharge.

The applicability and accuracy of recharge estimation techniques can also be compromised. Convectional rainfall is due in part to moisture recycling such that progressive rainfall over an area is accompanied by a decrease in the chloride concentration in the rainfall. Cyclones, on the other hand, bring chloride rich rainfall. Techniques that rely on the solute concentrations in rainfall such as the chloride mass balance method may yield erroneous values of the recharge estimate. The intensity of the rainfall event on the other hand may distort the estimates from methods that rely on the water holding capacity of the aquifer for the recharge estimate. For example, the water table fluctuation method may give an inflated hydrograph immediately after a heavy storm if preferential flow paths are present.

5.2.2 Rainfall characteristics from long record data

The long-record minimum monthly rainfall is 299 mm/a and the long term maximum is 1448 mm/a. Both were recorded at Beatrice station. The long-record average rainfall is 738 mm/a for Mubayira and 710 mm/a for Beatrice whilst the average coefficient of variation, CV, is 35% showing that the annual rainfall does not change significantly over the years. The minimum monthly rainfall for all the years considered is zero in the end months of the season, i.e., September, October, April and May whilst it is 3.8 mm/month for the mid season months of November to March suggesting that "it always rains in the rain season".

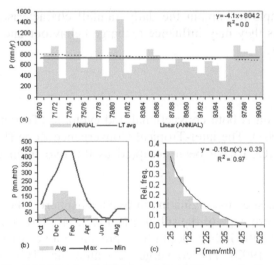

Fig. 5.4. Mubayira rainfall (a) annual total (b) seasonal pattern and (c) monthly depth frequency distribution.

The maximum-recorded monthly rainfall is 453 mm/month for Beatrice in the month of February. This value, like several other high rainfall months, does not coincide with the wettest year on record suggesting a rather erratic nature of rainfall pattern during a given season. It is therefore not possible to predict the annual total on the basis of the total rainfall for any given month.

Fig. 5.4 shows the annual total rainfall, the seasonal rainfall pattern and the monthly frequency distributions for the Mubayira rainfall station. Fig. 5.8 (a) shows that 18 years out of the 30 year record (60%) have rainfall below average. A normal frequency distribution positively skewed to the left fits the annual totals. Extremely wet years are less likely to occur compared to drier years. Thus if recharge occurs in direct proportion to rainfall more years are likely to have below average recharge.

Fig. 5.4 (b) shows that the average season starts in October, ends in April and follows a 'bell shaped' form with a peak in January. The behaviour is typical of tropical rainfall. Recharge is therefore expected to be greatest in the middle of the season between December and February. The maximum monthly rainfall can be up to three times the monthly average whilst the minimum monthly rainfall can fall by a similar magnitude. Inter annual variability in monthly rainfall is therefore high.

Fig. 5.4 (c) shows the frequency distribution of monthly rainfall for the rain months of October to April. The monthly rainfall follows a J type distribution suggesting a predominance of low rainfall months compared to high rainfall ones. A probability of exceedence for the monthly rainfall can be estimated by a logarithmic function of Equation 5.12.

$$P(P \geq p) \approx 0.33 - 0.15 \cdot \ln(p) \tag{5.12}$$

where p (mm/month) is the monthly rainfall. Equation 5.12 is only valid for the p range of 8 to 428 mm/month for which the probability is between 0 and 1, otherwise the monthly rainfall has to be considered an extreme event.

Attempts have been made to simplify the analysis of rainfall as well as bypass constraints due to data scarcity by determining relationships between daily and monthly rainfall for rainfall modeling purposes. For example, De Groen (2002) has shown that rainfall in the region follows a Markov[17] process and concluded that a two state Markov model can describe the relation between daily rainfall and monthly rainfall. The author proved that monthly interception, I_m can be estimated from a daily interception threshold, D, monthly rainfall, P_m and the number of rain days in a month, n_r as given in Equation 5.13.

$$I_m = P_m \left(1 - \exp\left(\frac{-D \cdot n_r}{P_m} \right) \right) \tag{5.13}$$

The effective monthly rainfall, $P_{eff,m}$ can then be estimated from Equation 5.14.

$$P_{eff,m} = P_m \exp\left(\frac{-D \cdot n_r}{P_m} \right) \tag{5.14}$$

---

[17] In this process the probability of an event in a time step depends on the state of the system in a previous time step.

5.2.3 Rainfall characteristics from the study period data

Since a long record of daily data for the Nyundo catchment does not exist analysis of rainfall is restricted to the data collected over the study period 1998/9 to 2000/1.

The rainfall tends to be convective but the monitored seasons suggest some cyclonic rainfall in January and February. Seasonal rainfall follows a 'bell shaped' feature with the highest rainfalls in January and/or February. The season lasts seven months from October to April.

A tabular comparison of rainfall data for all stations with respect to time shows that the rainfall usually occurs in storms and spells confined to definite periods within the month but rather random within the day. A rain period can be defined as a time interval in days that is bound by zero rainfall days on either end over which rainfall covers the gauged area. In this regard an average of 20 rain periods occurs per season in the gauged area. During the rain period, the rainfall events occur with a lag time of one to three days between stations. The lag time is defined here as the time that it takes for a similar rainfall event to occur at a nearby station.

The annual rainfall decreases in a north-westerly direction along the catchment with a gradient of approximately 2 mm/km. The increased rainfall to the southeast is most likely induced by the Mwenesi range of mountains which stretch southwards from this part of the catchment. The mountains cause some orographic rainfall in the upper parts of the catchment thereby increasing the rainfall in these parts compared to the lower parts of the catchment. It should be noted however, that the influence of the mountains is localized and the rainfall gradient as calculated is not uniform across the catchment.

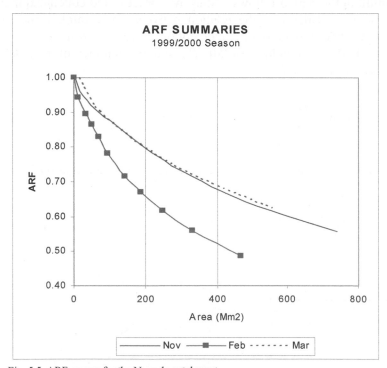

Fig. 5.5. ARF curves for the Nyundo catchment.

A depth area analysis (Shaw, 1994) of similar rainfall storms over the season yields the areal reduction factor (ARF) curve shown in Fig. 5.5. The ends of the season, November and March, suggest cyclonic influences on the rainfall whilst the middle part of the season, February, is dominated by convective rainfall, i.e., the gradient of the ARF curve is steeper.

Daily rainfall characteristics

A rain day is here defined as a calendar day with a measurable rainfall amount. In our case a value of 0.1 mm/d as recorded by the automatic gauge is considered as minimum rainfall that defines a rain day.

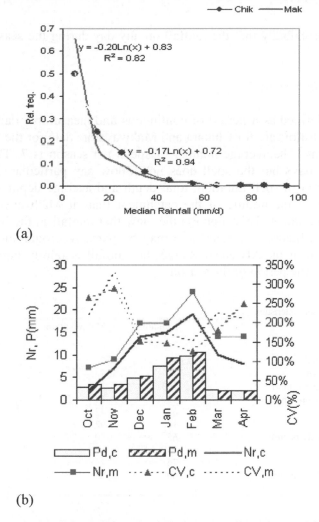

(a)

(b)

Fig. 5.6. Daily rainfall characteristics in the Nyundo catchment (a) frequency distribution, (b) depth, Pd, no of rain days, Nr, and CV.

NB: c and m denote Chikara and Makwavarara respectively.

Fig. 5.6 summarizes the characteristics of daily rainfall in the Nyundo catchment from the two main rainfall stations over the study period. The number of rain days per month varies between 2 to 7 days for the begin and end of the season and around 20 in the middle of the rain season. The number of rain days per season is around 105 and the season length can be taken to be 200 days, this being the number of days between the first and the last rain days of the season.

The average daily rainfall is about 6 mm/d but it fluctuates from a low figure of around 2 mm/d for the start and end of the season and a mean of around 10 mm/d for the middle of the season. The maximum daily rainfall is around 70 mm/d and occurs in the high rainfall month of February. The coefficient of variation for daily rainfall varies between 145% and 310% indicating highly variable rainfall occurrence.

The daily rainfall distribution is skewed to the left. Thus, the probability of little or no rain is a lot higher than the probability of having high rainfalls in the catchment. The probability distribution can be approximated from an exponential function of Equation 5.15 below.

$$P(P \geq p) \approx 2.2695 \cdot e^{-0.111p} \tag{5.15}$$

As an example, the probability that the rainfall on any day during the season will be above 30 mm is 0.08.

Wet spell and storm characteristics

A wet spell is here defined as a period of continuous and measurable daily rainfall. The descriptions that follow are for Chikara and Makwavarara data for the 1999/2000 and 2000/2001 seasons. The average number of spells per season is 7. The average spell length is seven days but the spell does not follow any particular shape. The coefficients of variation of spell duration, total depth and average depth show wide variations at Chikara in the middle of the catchment than at Makwavarara in the higher parts of the catchment. This supports the view that rainfall in the higher parts of the catchment is influenced by relief and may be partially orographic in nature whilst further away from the Mwenezi Range the rainfall is purely convective or cyclonic depending on the period of the season.

Table 5.2. Wet spell characteristics

Characteristic	Chikara	Makwavarara
No of spells per season	7	7
Average spell length (d)	7	7
Average total spell depth (mm)	103	101
Average spell daily depth (mm/d)	12	14
CV (duration)	64	42
CV (Total depth)	107	92
CV (Average spell daily depth)	65	60
CV (minimum P)	139	106
CV (Maximum P)	76	73
Spells as %age of seasonal total	60%	63%

Table 5.2 summarizes the spell characteristics for Chikara and Makwavarara. The table suggests that recharge most likely occurs during spells since daily rainfall exceeds potential evaporation and that variations in spell characteristics are mimicked in the recharge behaviour.

A storm is defined as a continuous rainfall event. For the analysis of storm rainfall in the catchment periods of continuous rainfall within an hourly time series were identified. The average duration of a storm is about 8 hours but the storm length can vary between 3 hours and 16 hours with a coefficient of variation of 52%. The total storm depth averages 46 mm with a range between 22 mm and 85 mm and a CV of 54%. The average storm intensity, as defined by the total storm depth divided by the storm duration, is about 9 mm/hr and ranges between 3 mm/hr and 12 mm/hr with a CV of 52%. The observed maximum intensity is 53 mm/hr but an average observed maximum intensity is around 37 mm/hr. The CV for maximum observed intensities is about 46%.

The occurrence of storms does not follow any detectable spatial pattern. Storms tend to be of shorter duration at the beginning and end of the season compared to the middle of the season.

The individual storm tends to dissipate its rain in either a bell shape or a positively skewed J shape suggesting some mixture of convective and cyclonic rainfall (De Laat, 1994).

5.2.4 Implications for groundwater recharge

The seasonal nature of rainfall means groundwater recharge can only occur in one half of the year and the other half is a discharge period. As such recharge estimation in such areas needs to consider this recharge-discharge relationship at an annual time step. The pattern of rainfall in the rain season means there are key months in which recharge is likely to occur. Those months with long wet spells favour groundwater recharge.

The limited spatial extent of storms means recharge will not be evenly spread in an area for which the areal extent is larger than that of the storm. Local groundwater flows will therefore influence the distribution of recharge water as water flows laterally from recharged areas to adjacent non-recharged areas within the same catchment.

The rainfall storm intensities are relatively higher than infiltration capacities for most soils. As such recharge at a point B may result from water that initially was runoff at the point of impact A some distance from point A. A mountain front type of recharge can therefore be expected when high intensity rainfall occurs.

5.3 Geology and Hydrogeology

This section presents the geological and geophysical investigations carried out in the catchment. The objective of the investigations was to develop a hydrogeological model that could be used for modeling the groundwater system.

Table 5.3. Geological formations in and around the Nyundo catchment.

Geological age/unit	Lithology
RECENT	Silty quartz feldspar sand
Period of erosion KALAHARI SANDS	Silcrete/silcrete rubble and sand
UPPER KAROO SYSTEM	Extrusive olivine basalts and limbergite
	Silty arkosic sandstone and sand
PRECAMBRIAN	Gneissic and massive granite
	Banded Ironstone (Jaspilite)

(Modified from Worst, 1962)

5.3.1 Geological investigations

Little or no work had been done to define the geology of the study area before this study. The Geological Survey Department contains three credible references on areas to the east and north of the catchment. Worst (1962) mapped the southern part of the area and reported in Bulletin No.54 on "The geology of the Mwanesi. Range and the adjoining country". Bulletin No 90 by Lister (1987) covered the area in passing where she discussed the erosion surface of the Beatrice triangle. Makuni and Simango (1987) produced a short report, No.52, on the area north of the Nyundo catchment.

Table 5.3 shows the main geological features in the wider study area. There are five main geological units. Starting from the base of the sequence upwards there are the Precambrian rocks, the Upper Karoo system, Kalahari sands and the recent rocks. Silty sandstone, sands and basalts represent the Karoo system that occurs extensively in Zimbabwe. Small patches of Precambrian granitic rocks are exposed in the upper Karoo sandstone.

A geological mapping exercise for the study area was carried out between June and December 2001. Among the prominent sites visited were six quarry sites, seven outcrops, five newly drilled holes, four gully slopes and several shallow wells. A total of thirty sites were considered. Local guides assisted in locating outcrops while soil colour was taken as a primary indicator of changes in geology (Barnes, 1995). Lithological boundaries were demarcated using a global positioning system (GPS) as topography alone was rendered inadequate by the undulating nature of the terrain. Rock samples were analyzed for petrography and mineralogy. Fig. 5.7 shows the deduced geological map.

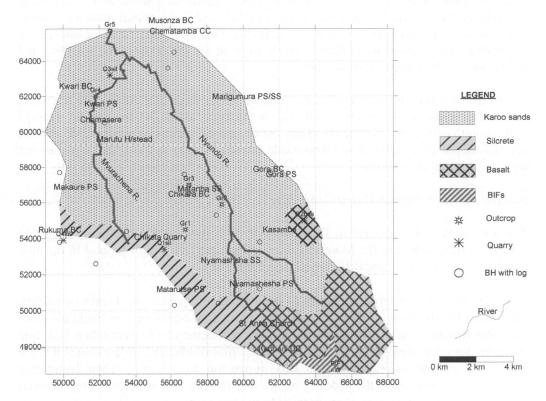

Fig. 5.7. Geological map for the Nyundo catchment.

Basement granitic rocks

Granitic rocks are exposed along stream channels in the central parts of the catchment. Elsewhere in the catchment the Karoo and Kalahari sands conceal these rocks. When exposed, the gneissic granite is pinkish in colour. The rock is medium to coarse grained in handspecimen.

Fig. 5.8. Joint filled with clay in granitic rock.

63

Fig. 5.8 shows that in the field, granitic rock consists of jointed and/or fractured gneissic granite occurring as a weathered regolith and fine-grained massive granite. The fractures trend roughly in a northsouth direction and some of the fractures are filled with clay. Groundwater flow is likely to be in the direction of the fractures and joints but will be impeded by the clay.

Quartz veins found in the gneissic granite trend in the same direction as the fractures. Pegmatite veins are relatively small with a thickness of 25-30 cm and an average strike length of four to five metres. They are composed mainly of quartz (3-8 cm) and alkali feldspar (5-10 cm). Some iron oxide coating is seen in the fracture planes but they are not that prominent.

Banded Iron Formations (BIFs) - Jaspilites

The BIF rock unit outcrops in the southeastern corner of the catchment area and extends southwards into the Mwanezi Range. The unit forms high ground in the area because of its hardness and differential weathering.

Jaspilite consists of alternating bands of haematite, and quartz with brown or reddish pink and whitish bands of chert respectively. Individual bands vary in thickness from very thin hairline size to 10 cm and occur more or less parallel to each other. Hair line fractures found in this unit are filled with some iron oxide. The quartz bands are 3-9 mm thick and occasionally exceed this dimension to 3.5 cm. When hammered strongly, it breaks along the bedding planes and usually fine striations, are observed on the haematite bands, which are parallel to the bandings.

The BIF is often in contact with a siliceous rock unit where the jaspilite rock fragments are found in the siliceous matrix (silcrete) forming a brecciated rock. This may mean that the contact is tectonic and the jaspilite is older than the silcrete.

Groundwater flow, as well as recharge, is through the fractures. However the small size of the fractures and the predominance of Iron oxides in the fractures may mean that flow quantities are less than for granitic formations. Flows increase were the rock mass is brecciated. As the formation is prevalent in the higher parts of the catchment surface runoff is likely to predominate over infiltration resulting in some form of monuntain front type of recharge at the interface of the BIF formation and the Karoo sand formations in the lower parts of the catchment.

Karoo sandstone and Sands

Fig. 5.9. Fractured and cracked sandstone.

The sandstone covers about 85% of the Nyundo Catchment surface area. The sandstone is creamish white in colour with a fine-grained texture (silty sandstone). Sedimentary structures are not very well developed although large cross beds with thin, 5 cm, laminations can be seen from careful observations. Observations at a gulley in silty sandstone reveal a well-sorted sandstone with shrinkage cracks and joints as shown in Fig. 5.9. Small bands of $CaCO_3$ (milky white in colour) are seen in the sandstone at this location without any specific direction.

The thickness of this unit is not uniform but varies from place to place. An average thickness of 33 m was indicated by Worst (1962) from elevation differences. An average thickness of about, 30 m was determined from borehole lithological logs obtained from the Groundwater Branch of the Department of Water Development.

Fine-grained silty sand grades into medium graded silty sandstone and a sandstone with carbonate nodules. The nodules have a blackish colour as a result of chemical weathering. The carbonate nodules have been affected by preferential chemical weathering leaving behind some small spherical hollows of about 4 to 12 cm in diameter. The rock is not well consolidated possibly because the cementing material was leached out or the sediments were never consolidated at all.

Aeolian and fluvial depositional environments can give rise to this type of rock unit. The loose sands present are typical of desert (i.e. arid to semi arid) conditions. Well sorted sediments and fine grain sizes suggest aeolian depositional environments. Textural and compositional maturity variations in a rock sample are typical of a fluvial environment. The presence of calcrete and lack of fossils supports arid fluvial conditions but fluvial sediments deposited under semi-arid climates are usually red in colour (Tucker, 1991).

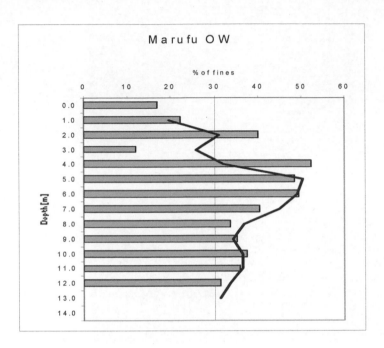

Fig. 5.10. Vertical distribution of fines in a Karoo sand profile.

Progressive leaching results in the accumulation of fines at depth. Fig. 5.10 shows that the percentage of fines (clays and silts) increases to a maximum of 50% by weight at 5 to 6 m depth before stabilizing around 30%. This behaviour can be attributed to diffuse groundwater recharge which tends to leach out fine particles in suspension to the bottom of the minimum depth of the water table around 5 to 6 m below ground surface.

Typical soil horizons consist of 50 - 100 cm of fine sand, sometimes without clay, abruptly overlaying a mottled loam or sandy clay with a weak blocky or prismatic structure and well-developed clay and sand cutans. The overlying sandy material appears to meet the requirements of an Albic E horizon, that is one that has lost clay and iron oxides and whose colour comes from the uncoated sand particles, which tend to be very pale brown when dry.

The Karoo sandstone and sands form the main aquifer in the Nyundo area. The fine to coarse grained, loose nature of the formation gives rise to reasonable permeabilities and high effective porosities and specific yields. Water holding and transmission capacities are therefore higher than for the other formations. Infiltration capacity is also higher which condition favours groundwater recharge.

Karoo Basalts

Karoo basalts cover about 10% of the study area and outcrop mainly in the southeast corner of the catchment. Other basalt outcrops occur in the eastern parts of catchment. Basalts exposed in the southeast have some light to pale green filled vesicles when compared to those further north of the catchment. This shows that there are different varieties of the Karoo basalts in the catchment. The outcrops in the southern margin are quite different mineralogically to the outcrop found east. The basalt is weathered into globules of 5 to 10 mm in diameter, while the weathered surface is whitish green in colour. The basalts overlie the sandstone at many localities but it has also been reported that the basalts overlie the granites in some areas adjoining the study area (Worst, 1962). The same author used elevation differences to estimate the original thickness to exceed 117 m.

The Karoo basalts and BIFs give rise to red clayey silts which tend to have shallow depths ranging between 0.5 m and 1.0 m overlying a thicker heavily weathered granular basaltic/BIF regolith at most locations.

Silcrete

This rock formation is only about 0.5 to 1.2 m thick and occurs on top of Karoo sandstone and siltstone. The unit is found in the southern margin of the study area and outcrops at the borders of the Karoo basalts. In other parts of the catchment it occurs as reddish brown glassy and/or translucent glassy rubble. The unit usually gives a dull red colour to the soil.

Kalahari sands

These sands are found as small patches in the study area. It is very difficult to distinguish between the Karoo sands and the Kalahari sands in the field. The sand is reddish in colour and is distinguished from the Karoo sand that is more white in colour. The grains are well rounded indicating an aeolian transportation agent. The thickness of the sands is about 4 to 8 m from borehole lithological logs.

5.3.2 Geophysical investigations

The electrical resistivity method has been selected for this study because of (1) its ease of application, (2) of the superior options it offers to investigate the subsurface properties, (3) its fully automated data processing facilities (several packages are on offer for resistivity interpretations), and (4) the wide application that the method already enjoys.

In crystalline basement aquifers the geo-electrical base is revealed by a forty-five degree rising arm after the minimum curve turning point (Lloyd, 1999). For most

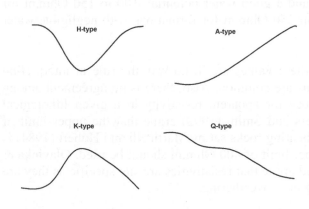

Fig. 5.11. Types of three layer master curves (After Van der Moot, 2001).

practical cases resistivity interpretations in crystalline basement aquifers are adequately described by three layer master curves for which there are four basic curve types (Nonner, 1995; Lloyd, 1999) as shown in Fig. 5.11. The H-type curve indicates low resistivity material underlain by hard crystalline rock and is very common in the crystalline basement aquifers. The A-type curve indicates increasing resistivity and can be typical for outcrops of crystalline or intrusive rocks whereby the water content of the materials decreases with depth. The K-type indicates a high resistivity formation underlain by lower resistivity material whilst the Q-type is prevalent in formations for which resistivity decreases with depth such as salt water intruded areas.

There are no simple straight rules on resistivity values for crystalline basement aquifers. Attempts have been made to group the resistivity readings from crystalline basement aquifers and thereby interpret geological formations and their thicknesses from apparent resistivity readings.

Martinelli and Hubert (1985) simplified and generalized resistivity readings for granites, gneisses and paragneisses in the African and Post-African erosional surfaces as shown in Table 5.4. The table indicates that resistivity readings generally increase with a decrease in weathering and water bearing potential of igneous rock.

Table 5.4. Resistivity ranges for granites, gneisses and paragneisses in the African and Post-African surfaces (After Martinelli & Hubert, 1985).

LITHOLOGICAL SUCCESSION	Resistivity range (Ohm.m)	AQUIFER CHARACTERISTICS
Various top soils	80 – 1000+	Superficial layer, dry. Resistivity depends on clay/sand ratio.
Highly weathered granite, gneisses and paragneiss	20 - 50	Main water bearing horizon when thickness of overburden exceeds 25 m.
Weathered granite, gneisses and paragneiss	50 - 100	Important water bearing horizon when thickness of overburden exceeds 25 m.
Partly weathered granite, gneisses and paragneiss	100 - 250	Moderate to important aquifer.
Poorly weathered granite, gneisses and paragneiss	150 - 250	Marginal aquifer.
Unweathered and unfractured granite, gneiss and paragneiss	500 – 2000+	Hydrogeological bedrock

Note: Interpretations are based on electrical resistivity data only.

Wright (1992) further simplified apparent resistivity readings and gave resistivity ranges with respect to the aquifer potential for the layered regolith. He suggested a reading less than 20 Ohm.m for clays with limited potential or saline water, 20 to 100 Ohm.m for optimum weathering and a good water potential, 100 to 150 Ohm.m for medium conditions and larger than 250 Ohm.m for formations with negligible water potential.

It should be noted however that these values are in no way the rule in interpreting resistivity values as local variations are common. Also, there is no agreement among researchers on the indicative values for apparent resistivity in a given lithological formation. For example, Carruthers and Smith (1992) argue that the upper limit of 150 Ohm.m for potentially water bearing rocks set by Martinelli and Hubert (1984) is rather low and suggest that an upper limit of 400 Ohm.m should be used. Olayinka & Sogbetum (2002) on the other hand argue that resistivities are site specific as they are determined by mineralogy and degree of weathering.

In order to link the resistivity readings to lithological formations and thus enable cross and longitudinal sections of the aquifer system to be drawn a calibration table can be prepared (Van der Moot, 2001; Nonner, 1995). The calibration table assigns resistivity values to lithological types taking into consideration the degree of saturation and the salinity of the aquifer water. Table 5.5 summarizes the lithologies, their calibration resistivities and indicative depths for the Nyundo catchment as derived from field data and literature guidelines as discussed above.

The resistivity survey itself was conducted as much as possible along traverse lines at least 2 km apart and at a station spacing of at least 500 m. Each measurement was completed to the bedrock of the crystalline basement . The positions of measurement points were established with a Garmin 12 GPS with an accuracy of 10 m. Topographic surface levels were initially based on the 1:50 000 topographical map (SGO, 1990) but later verified with field leveling and differential GPS measurements where feasible.

Borehole logs were required to calibrate the resistivity measurements and estimate lithological distributions and layer thicknesses. A total of forty (40) borehole data sets covering the whole catchment were collected from the groundwater database of the Zimbabwe National Water Authority (ZINWA), District Development Fund (DDF) records and private borehole drillers. The borehole data was processed using the United Nations groundwater database management program, UN-GWW[18].

On the borehole logs, lithologies were grouped to simplify the relation of geology to resistivity values as well as the construction of the subsequent conceptual model (Van der Moot, 2001; Anderson & Woessner, 1991). The lithologies in the Nyundo could be grouped into four structural zones; the soil zone, the decomposed/weathered zone, the fractured zone and the fresh rock zone as summarized in Table 5.5.

Table 5.5. Lithological classifications and simplifications for the Nyundo catchment.

GEOLOGICAL ZONE	CONSTITUENT LITHOLOGIES	RESISTIVITY RANGE (ohm.m)	AVERAGE DEPTH (m)
Soil zone	Karoo sand, silt, clayey silt, clayey loam	5 – 25	0.2 – 2
Decomposed or weathered zone	Karoo sandstone, siltstone.	20 – 60	2 – 30
	Granular basalt; gravelly granite, gneises, dolerites.	50 - 250	8 - 40
Fractured zone	Granite, gneiss, dolerite	200 - 550	8 - 40
Fresh rock	Granite, gneiss, dolerite	500 - 1500	8 -40

Resistivity findings

Fig. 5.11 gives examples of the main types of resistivity curves obtained in the study area. The catchment is dominated by the H-type curve which occurs in more than 75% of the surveyed areas. Such curves occur in areas where groundwater is present below a drier top soil and above the harder crystalline rock. This result is consistent with most findings in crystalline basement or hard rock aquifers (Lloyd, 1999). The distribution of the curves is fairly uniform over the catchment.

---

[18] This is a public domain package representing a relational database in the sense that a borehole is given a unique identification which relates all well data like lithological formations and depths, water levels, start and end of casing, water chemistry, etc.

The next important curves are those resembling mostly the A-type category and are registered in about 20% of the measurements. These curves occur in areas with a thin soil cover or at outcrops where groundwater is hardly or not present above the solid crystalline formation. The A-type curve was observed mainly in the upper southeastern parts of the catchment, at some isolated patches within the catchment and around the catchment outlet. The occurrence of Q-type curves, at about 5%, was negligible and restricted to a small patch towards the northwestern end of the catchment.

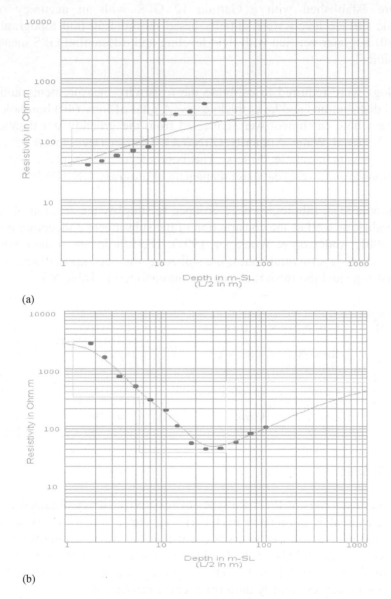

(a)

(b)

Fig. 5.11. Typical resistivity curve plots (GEWIN) from the Nyundo catchment.

Some interpreted geological cross sections from the area are shown in Fig. 5.12. It should be pointed out however that, in some cases it is difficult to separate the Karoo system from the weathered regolith layers because both are thin and have almost the same range of apparent resistivity (principle of suppression and equivalence).

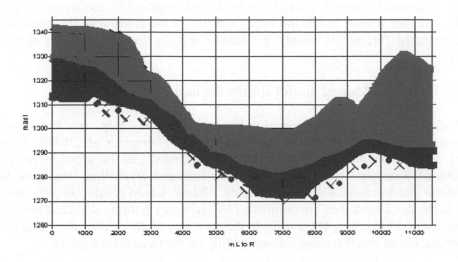

(a)

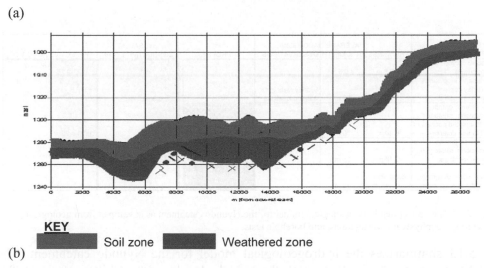

KEY

Soil zone   Weathered zone

(b)

Fig. 5.12. Some cross sections deduced from resistivity measurements, (a) perpendicular to Nyundo river flow (section B-B in Fig. 5.2) and (b) parallel to Nyundo river flow (Section 2-2 in Fig.5.2). [Sketch not to scale].

Resistivity profiles parallel to the main river channel suggests that the thickness of the weathered zone decreases as one moves upstream. The bottom boundary for the soil zone mimics the topography of the ground surface to some degree suggesting a depositional nature, suspiciously aeolian, of the Kalahari/Karoo sediments (Worst, 1962). The soil zone thickens in the middle to lower parts of the catchment. The zone also is thin along the river channel supporting the theory of erosion and transportation along the river course (Wright, 1992). Indeed the formation of gulleys are common along the river channels.

The weathered zone behaves similarly to the soil zone in terms of thickness vis a viz the river orientation. However, the bottom boundary does not mimic the topographic surface, at some points even going anti-slope. This is to be expected as weathering is likely to be highest at those points were fractures are prevalent and water, the weathering agent, could be trapped (Taylor & Howard, 2000). Consequently, the highest points on the basement rock are the least fractured and the thickness of the regolith is at its minimum. The weathered zone is either very thin or non-existent were terrain slopes are above 1%. This is likely so because hardly any weathering occurs on slopes since rainwater has a short residence time on the slopes.

71

Alternatively when the weathering does occur erosion quickly leads to transportation of the weathered material leaving the fresh rock exposed to further erosion (Pitts, 1984). Another reason is that the upper reaches of the catchment are dominated by more resistant formations mainly the basalts and the banded ironstone formations (BIF).

5.3.5 Hydrogeological model for the Nyundo catchment

Geological mapping and geophysical investigations help to formulate the hydrogeological model for a groundwater system (Lerner et al.., 1990, Simmers, 1997). For the Nyundo catchment it can be concluded that the groundwater system consists of three aquifers; weathered and or fractured crystalline, weathered Karoo sandstone and unconsolidated Karoo sands. Though Karoo basalts, silcrete and BIFs exist in the catchment they are dominated by secondary porosity and their areal extent and distribution is negligible relative to the total size of the catchment. Consequently, the discussion of this discourse focuses mainly on the three main aquifers.

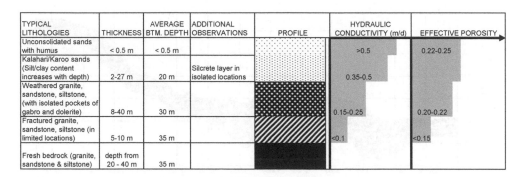

TYPICAL LITHOLOGIES	THICKNESS	AVERAGE BTM. DEPTH	ADDITIONAL OBSERVATIONS	PROFILE	HYDRAULIC CONDUCTIVITY (m/d)	EFFECTIVE POROSITY
Unconsolidated sands with humus	< 0.5 m	< 0.5 m			>0.5	0.22-0.25
Kalahari/Karoo sands (Silt/clay content increases with depth)	2-27 m	20 m	Silcrete layer in isolated locations		0.35-0.5	
Weathered granite, sandstone, siltstone, (with isolated pockets of gabro and dolerite)	8-40 m	30 m			0.15-0.25	0.20-0.22
Fractured granite, sandstone, siltstone (in limited locations)	5-10 m	35 m			<0.1	<0.15
Fresh bedrock (granite, sandstone & siltstone)	depth from 20 - 40 m	35 m				

Fig. 5.13 The conceptual hydrogeological model for the Nyundo catchment as developed from geological mapping, geophysical investigations, and borehole tests.

Fig. 5.13 summarizes the hydrogeological model for the Nyundo catchment. The model bears a strong similarity with the models developed by Chilton et al., (1984) and Forster (1984) from data collected in Malawi. The Nyundo catchment is therefore a typical crystalline basement aquifer. Its only unique feature is the prominence of the Karoo sands and sandstones.

The compound[19] aquifer thickness, from the topographic surface to the fresh crystalline basement varies from as thin as 5 m to as thick as 40 m. A larger thickness is possible as highlighted from geophysical measurements but the extent of such thicknesses is doubtful given that the crystalline base is fairly undulating.

The thickness of the unconsolidated layer ranges between 2 and 30 m with an average of 20 m. Porosities are high between 0.35 and 0.45. The layer properties are reasonably homogenous[20] and isotropic[21] given the uniform nature of the sands.

---

[19] This is the total thickness of the aquifer when the different layers are not considered separately.
[20] The aquifer properties do not change from place to place.
[21] Properties remain the same in both x and y directions.

Permeability is the highest for all aquifer materials with values between 0.05 and 0.2 m/d. Transmissivities are not expected to exceed 5 m^2/d (Martinelli, 1999).

The thickness of the weathered zone of either granite or Karoo sandstone varies from 6 m to 37 m with an average value of 23 m and a coefficient of variation of 38%. Porosity is heavily reduced compared to the unconsolidated sands primarily due to finer materials, mostly clay, trapped in the pore spaces of the coarser grained materials (Brassington, 1997). Values within the range 0.1 to 0.15 are possible. Permeabilities are also lower and fluctuate between 0.03 and 0.4 m/d, the Karoo sandstone being on the lower side (McFarlane, 1992).

Isolated fractured zones exist below the weathered zone in the central parts of the catchment. Such zones have a thickness between 10 m and 13 m.

5.4   Vegetation

Vegetation affects the hydrological cycle in a number of ways. The presence or absence of trees, for example can influence river flows (Mumeca, 1986; Nhidza, 1999) and groundwater recharge (Sandström, 1995).

Data for the catchment was collected from 1:25 000 aerial photos (SGO, 1984), 1:50 000 topographic maps (SGO, 1990). Fieldwork essentially focused on confirmation and detailing of the data from the aerial photos and topographical maps. Tree density counts, species naming and mapping of the areal extent of any homogeneous tree species was carried out.

The data and field visits revealed that between 60% and 80% of the total area is used for dry land farming with the remaining 20% to 40% taken up by pastureland (20%), marshes (4%), indigenous forestry (14%) and gum tree plantations (2%). The vegetation is mainly Savanna. Tree density counts of 200 to 300 trees per hectare are the mode in the bushy areas and densities of 500 to 1 000 trees per hectare are prevalent in the grasslands. Fig.5.14 gives a deduced vegetation map for the area.

An important characteristic feature of the catchment is that most of vegetation sheds leaves or dries out in the dry season such that bare soil covers approximately 70% to 100% of the total area during this period.

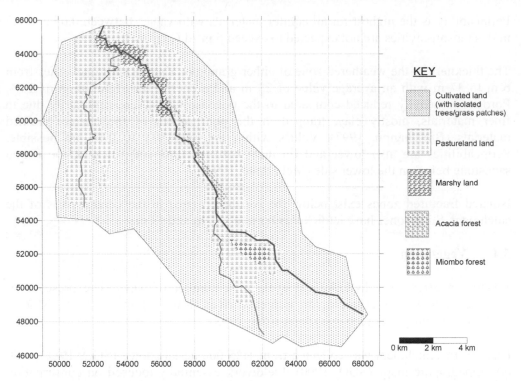

KEY

- Cultivated land (with isolated trees/grass patches)
- Pastureland land
- Marshy land
- Acacia forest
- Miombo forest

0 km   2 km   4 km

Fig. 5.14. Vegetation cover map for the Nyundo catchment.

5.5   Summary and conclusions

This chapter has shown that the rainfall in the Nyundo catchment is seasonal and follows the 'bell shape' distribution in the season. Storms are of limited spatial extent and short duration. Wet spells can last as long as seven days on average. It is during the peak rainfall months and during the wet spells that recharge is likely to occur.

The geology of the area is dominated by Karoo sandstone and sands underlain by a weathered granitic base forming a two layer crystalline basement aquifer. Aquifer layer thicknesses vary between 2 m and 30 m. Natural vegetation is typically Savanna whilst over 60% of the land area is used for dryland farming. The recharge process is influenced by man's activities.

CHAPTER SIX

6. CATCHMENT WATER BALANCES AND WATER BALANCE MODEL

An analysis of the catchment water balance can reveal the nature of the fluxes in the catchment, their absolute and relative magnitudes as well as their relationships to each other (Cook, et al., 1998). Taken at short time scales, say monthly, the catchment water balance can indicate when direct groundwater recharge is likely to occur (Da Silva, 1999).

The objectives of this section are (1) to quantify catchment fluxes and storage at annual, seasonal and monthly time steps, (2) to deduce basic linear relationships for the different water balance components as functions of rainfall, and (3) highlight the difficulties associated with groundwater recharge estimation using a catchment water balance approach.

6.1 The water balance for the Upper Mupfure catchment

The absence of long record[22] series of data on the Nyundo catchment necessitated the evaluation of a long record water balance for the Upper Mupfure. It is the argument in this thesis that since the two catchments have similar physical characteristics, are about 5 km apart and lie within the same Mupfure catchment their hydrological behavior is similar. Table 6.1 gives a comparison of the physical characteristics of the two catchments. The similarities in land use, soil types, geology and terrain make the two catchments reasonably equivalent.

Caution needs to be exercised however. Farming practices influence the hydrological behavior of a catchment (Mumeca, 1986). In large commercial farming areas of Zimbabwe more land is left fallow and vegetated areas tend to be larger such that both surface runoff and baseflow may be reduced to compensate for increased interception and transpiration. The stream discharge in the Upper Mupfure may therefore be reduced compared to that from the Nyundo catchment.

---

[22] Long record in this sense refers to data series covering several years whilst study period refers to the data collected over a few years during the course of the study.

Table 6.1. Physical similarities between the Upper Mupfure and Nyundo catchments.

	Upper Mupfure	Nyundo
Area (Mm^2)	1,250	180
Human settlement (%area)	60% 'commercial farms' 10% 'resettlement' 30% 'communal'	100% communal
Vegetation cover	40% crop cultivation 30% woodland/forestry 30% grassland/pasture	50% crop cultivation 10% woodland/forestry 40% grassland/pasture
Terrain slope	Gentle	Gentle
Soils	80% Karoo sands 20% red/black clay/silt	85% Karoo sands 15% red/black clay/silt
Geology	12% granitic outcrops, 11% *BIFs, 12% basalts, rest Karoo sand.	12% granitic outcrops, 11% *BIFs, 12% basalts, rest Karoo sands.
Drainage pattern	Dendritic	Dendritic

6.1.2 Upper Mupfure long period water balance

In the absence of data on all the components of a water balance, it is possible to evaluate a long record water balance at a monthly time step using only two parameters: usually rainfall and stream discharge. Some authors have applied this approach to study the water balance of catchments in varied climatological, hydrological and physiological settings. For example, Xiong & Guo (1998) argue that the interrelation between rainfall, runoff and evaporation on a monthly time step is close since irregular short time effects are smoothened out resulting in simpler hydrological models with fewer parameters. Najjar (1999) analyzed rainfall, temperature and stream discharge data to develop a simple statistical and water balance model for the catchment outlet stream discharge for the Susquehanna River basin.

A simplified water balance model, similar to the approach of Najjar (1999), has been used to evaluate the Upper Mupfure catchment water balance using rainfall and discharge data. The data used for the water balance computations is from the Beatrice rainfall station and the C70 river flow gauging station on the Upper Mupfure River. Fourteen years of data were used for the analysis.

The annual catchment water balance can be written as:

$$\frac{\mathrm{d}S}{\mathrm{d}t} = P - Q - E \qquad\qquad (6.1)$$

Where all the terms are as described in Section 3.2 of this thesis. Since there is no accumulation of water in the catchment over a long period of time, the long-term change in storage is negligible such that the long-term water balance can be written as:

$$E \cong P - Q \qquad\qquad (6.2)$$

All parameters represent long record average values. Further, since recharge water partly ends up in surface channels and stream gauges measure both flood flow and baseflow, the discharge term in Equations 6.1 and 6.2 represents the total discharge from the catchment (Roberts & Harding, 1996). The term evaporation, E, in this equation, being the residual of the water balance represents all actual fluxes other than the rainfall and stream discharge. As such, it incorporates evaporation from interception, open water, bare soil and transpiration, the portion of spring discharge that does not emerge as baseflow and well abstractions. The latter two fluxes can be considered relatively small (as shall be evident in later discussions) such that the term E can safely represent the evaporative fluxes, i.e., evaporation from interception, open water, bare soil and transpiration.

Table 6.2. The long record average water balance for the Upper Mupfure catchment.

	Oct	Nov	Dec	Jan	Feb	Mar	Apr	May	Jun	Jul	Aug	Sep	Sum
P	38.8	89.8	140.5	147.9	104.8	55.3	26.9	3.3	0.0	0.7	0.9	6.6	615.7
Q	1.9	4.5	7.0	7.4	5.2	2.8	1.3	0.2	0.0	0.0	0.0	0.3	30.8
E	36.9	85.3	133.5	140.5	99.6	52.5	25.6	3.1	0.0	0.7	0.9	6.3	584.9

The data in Table 6.2 confirms the widely held view that evaporation is the major flux in semi-arid tropical catchments and that its understanding is crucial in determining groundwater recharge in catchments like the Upper Mupfure. At an annual time step evaporation accounts for 95% of the rainfall whilst stream discharge accounts for only 5% of rainfall and is 19 times less than evaporation.

Hydrological relationships from the Upper Mupfure long record water balance

Plots of the long record rainfall and discharge reveal an 85% correlation between rainfall and stream discharge. The annual stream discharge can therefore be estimated directly from annual rainfall. Fig. 6.1 shows that the annual stream discharge fits a linear relationship with annual rainfall. A threshold annual rainfall depth of 400 mm/a exists below which stream discharge is negligible. Butterworth et al., (1999a) made similar observations from a field scale study in the semi arid Romwe catchment in south-eastern Zimbabwe.

A simple regression equation for annual discharge as a function of rainfall can be deduced from the scatter plot of Fig. 6.1. Thus the annual stream discharge for the Upper Mupfure can be estimated from annual rainfall by Equation 6.3.

$$Q = 0.2 \cdot (P - 400) \tag{6.3}$$

A similar expression is derived for evaporation by substituting Equation 6.3 into Equation 6.2 to obtain Equation 6.4.

$$E = 0.8 \cdot (P + 100) \tag{6.4}$$

Equation 6.3 shows that annual runoff is below 20% of total annual rainfall and that a rainfall threshold of 400 mm/a has to be exceeded before stream discharge becomes significant.

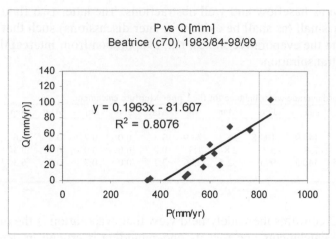

Fig. 6.1. The relation between annual rainfall and annual stream discharge in the Upper Mupfure catchment.

If the long record mean annual rainfall of 730 mm/a is substituted into Equations 6.3 and 6.4 the "runoff coefficient" is 9% making the "evaporation coefficient" 91%. Further, if a baseflow index (BFI) of 0.6 is assumed the long record water balance would suggest a "recharge coefficient" of about 6% of annual rainfall for the Upper Mupfure catchment.

In reality however, hydrological processes are non-linear and equations are merely indicative of the expected annual values of discharge and evaporation. To better account for non-linearity the above equations are treated as minima/maxima functions that define threshold values for the discharge and evaporation.

$$Q = 0.2 \cdot \max(0, P - 400) \tag{6.5}$$

$$E = \min\{0.8 \cdot (P + 100), P\} \tag{6.6}$$

Such functions shall be used in the Nyundo daily water balance model.

6.2   Estimates of water balance components

The water balance unit for this study is the water depth in mm per time unit.

6.2.1 Rainfall

In the Nyondo catchment, two rainfall station have been used. (See also Section 5.2). The Chikara station data has been used as the catchment average daily rainfall because of its central location within the catchment.

6.2.2 Interception

The models for estimating interception are not yet well developed nor are their results reliable. Models such as Rutter et al. (1971) which use the water balance of the interception storage were proven inadequate in tropical areas were rainfall intensity is higher, storm durations short, evaporation high and raindrop sizes bigger. Models have however been developed for the tropical areas but these are based on probability theory rather than the actual behaviour of the rainfall itself (Calder et al., 1986; Milly & Eagleson, 1982).

For this study interception has been defined as the fast (day time scale) evaporative feed back into the atmosphere of the rainfall that does not reach the root zone or the drainage system (Savenije, 2004). Simple assumptions about interception have been made. (1) The daily interception storage has a maximum value and thus the interception is limited and (2) interception takes precedence over any other processes. Thus the net daily rainfall, P_{net} (mm/d) is the difference between the observed rainfall and a daily interception threshold, I_p. In this case the actual interception is the minimum of actual rainfall and the interception threshold.

6.2.3 Potential transpiration, T_p

The potential daily transpiration has been calculated (De Laat, 2001) using the Penman equation with long period data from the Belvedere meteorogical station in Harare. This was the nearest station with all relevant data. The rationale for using the Belvedere data is that the coefficient of variation of potential evaporation in space is low. As such a single open water value can be used over a relatively large area provided the climatic conditions remain the same. The Belvedere meteorological station is about 60 km from the Chikara rainfall and evaporation station and about 100 m higher in altitude.

The open water evaporation, E_o, has been estimated from the Penman equation (De Laat, 2001).

$$E_o = \frac{C}{\rho L} \cdot \frac{s \cdot R_N + c_p \cdot \rho_a \cdot (e_a - e_d)/r_a}{s + \gamma} \qquad (6.7)$$

E_o is the open water evaporation (mm/d), C is a constant to convert m/s to mm/d ($C=86.4*10^6$), R_N is the net radiation at the earth's surface (W/m^2), L is the latent heat of vaporization ($L = 2.45\times10^6$ J/kg), ρ is the density of water (kg/m^3), s is the slope of the temperature-saturation vapour pressure curve (kPa/ $^\circ$K), c_p specific heat of air at constant pressure ($c_p = 1004.6$ J/kg/ $^\circ$K), ρ_a is the density of air ($\rho_a = 1.2047$ kg/m^3 at sea level), e_d is the actual vapour pressure of the air at 2 m height in kPa, e_a is the saturation vapour pressure of the air temperature at 2 m height in kPa, γ is the psychrometric constant ($\gamma = 0.067$ kPa/ $^\circ$K at sea level) and r_a is the aerodynamic resistance in s/m.

Potential transpiration is difficult to estimate. In this study the two-stage approach has been adopted. The method is relatively cheap and well practiced in the region. It has often given reasonable results in water balance studies. For example, Everson, (2001) showed that the estimates of evaporation from the Penman approach can differ by between 5 and 20% from estimates based on the Bowen ratio energy balance approach. The potential transpiration, T_p, has been estimated from Equation 6.8.

$$T_p = k_c \cdot E_o \tag{6.8}$$

where k_c [-] is a crop factor (Doorenbos & Pruitt, 1992). The crop factors and potential evaporation values used in the study are shown in Table 6.3.

Table 6.3. Crop factors and daily potential evaporation rates (mm/d).

Month	Oct	Nov	Dec	Jan	Feb	Mar	Apr	May	Jun	Jul	Aug	Sep
*Ref. Evapo, E_o	5.2	5.3	5.1	4.5	4.3	4.2	3.5	3.0	2.6	3.0	4.0	4.7
CROP FACTORS												
Per. veg.	0.85	0.85	0.85	0.85	0.85	0.85	0.85	0.85	0.85	0.9	0.9	0.85
Miombo	0.95	1.05	1.15	1.15	1.15	1.10	0.90	0.85	0.0	0.0	0.0	0.85
Acacia	0.85	0.85	0.85	0.85	0.85	0.85	0.85	0.85	0.0	0.0	0.0	0.85
Pasture	0.50	1.05	1.15	1.15	1.15	1.15	1.15	1.05	1.05	1.05	0.5	0.50
Wetland	0.75	0.75	0.90	0.90	0.90	0.90	0.90	0.90	0.75	0.75	0.75	0.75
Maize	0.0	0.0	0.65	0.65	0.65	1.05	1.05	0.30	0.30	0.0	0.0	0.0

*ref. Evapo = reference evaporation, Per. Veg. = perennial vegetation, Miombo = miombo woodland, Acacia = acacia woodland, pasture = pasture savanna.

Evaporation pan readings were also taken during the study and T_p estimated from Equation 6.9.

$$T_p = k_c \cdot k_{pan} \cdot E_{pan} \tag{6.9}$$

where k_c is a crop factor as above, k_{pan} [-] is the pan coefficient to convert pan readings into potential evaporation readings, E_{pan} [-] is the recorded pan evaporation. A pan coefficient of 0.8 has been used for initial estimates in line with standard practice by the national hydrology department.

To account for the spatial heterogeneity in a lumped parameter model a weighted average based on land use types has been used as given in Equation 6.10.

$$T_p = \sum_{i=1}^{n} T_{pi} \cdot a_i \tag{6.10}$$

where, T_p [-] is the average potential transpiration over the entire area under consideration (mm/d), T_{pi} [-] is the potential transpiration over a sub area with land use type i (mm/d), a_i is the areal fraction of landuse type i in the entire area and n is the number of landuse types considered. (See also Section 5.4). Table 6.4 gives the catchment monthly maximum transpiration rates.

Table 6.4. Monthly maximum transpiration (mm/d).

	Oct	Nov	Dec	Jan	Feb	Mar	Apr	May	Jun	Jul	Aug	Sept	Avg
Rate	2.7	3.2	3.3	2.9	3.0	3.3	2.8	1.8	1.2	1.4	1.2	2.0	2.4

6.2.4 Stream discharge, Q_o

The daily catchment stream discharge was measured at the catchment outlet and processed as outlined in Section 5.1. The observed stream discharge has been taken as the response parameter in calibrating the water balance model.

6.2.5 Changes in saturated storage, dS

The storage changes in the saturated zone can be estimated from either the flux flows into and out of the storage or from level changes in the host medium.

In the first method the storage change, ΔS_g [L] is calculated as the resultant of inflows, I [L] and outflows, O, [L] into the storage over the balance period, Δt [T].

$$\Delta S_g = (I - O) \cdot \Delta t \tag{6.11}$$

If the inflows and outflows are functions of time the total change in storage over a series of time steps can be mathematically expressed as:

$$\int_{St_o}^{S_t} dS_g = \int_o^t [I(t) - O(t)] dt \tag{6.12}$$

where the subscripts t_o and t denote the times at the beginning and end of the balance period.

In the second approach the storage change is estimated from the hydraulic properties of the storage medium. The concept of the specific yield is applied. This can be defined as the amount of water that a phreatic aquifer can release or store per unit change in groundwater head. The storage change in the balance period is given as the product of the specific yield, S_y [L/L], and the total change in head, Δh [L].

$$\Delta S_g = S_y \cdot \Delta h \tag{6.13}$$

The total change in storage after a series of events within the balance period can be mathematically expressed as:

$$\int_{St_o}^{S_t} dS_g = S_y \cdot \int_{h_0}^{h_t} dh \tag{6.14}$$

Since the storage change is the same Equations 6.12 and 6.14 can be combined.

$$S_y \cdot \int_{h_0}^{h_t} dh = \int_o^t [I(t) - O(t)] dt \tag{6.15}$$

Equation 6.15 offers a way for calibrating a water balance model on the basis of groundwater level fluctuations. This calibration approach is used in the daily water balance model of this study and in more established 1-D models, e.g., the EARTH groundwater recharge estimation model (Van der Lee & Gehrels, 1990).

6.3   The Nyundo catchment daily water balance model

The objective of the water balance study was to determine the magnitudes of stocks and fluxes in the catchment and their relation to rainfall over time. Of particular interest was the estimation of the occurrence and magnitude of recharge. A correct elaboration of the catchment water balance gives an indication of the likely hydrological relationships in a catchment as well as helps identify correct and relevant water resources management strategies.

A lumped parameter approach has been used in analyzing the Nyundo catchment water balance. The overall catchment water balance has been discussed in Section 3.2 and can be given as:

$$\frac{\mathrm{d}S}{\mathrm{d}t} = P - Q - T - I \qquad (6.16)$$

where all the symbols are as defined in Section 3.2. The interception, I, in this case represents the "white" water fluxes including the open water evaporation, direct bare soil evaporation and canopy interception. The transpiration term, T, represents the "green" water fluxes and includes both transpiration from the unsaturated zone and from the saturated zone. The stream discharge term, Q, represents all the "blue" water fluxes from the catchment. The discharge term combines baseflow, interflow, spring discharges and human abstractions. The storage term, $\mathrm{d}S/\mathrm{d}t$ represents the total change of storage in the catchment. It includes changes in the surface, unsaturated and saturated zones.

6.3.1 Rationale and basis of the daily water balance approach

This section describes the development of a daily lumped conceptual water balance model for the Nyundo catchment. The daily water balance model traces the fate of a rainfall drop through all moisture storage zones until it exits the catchment at the catchment outlet.

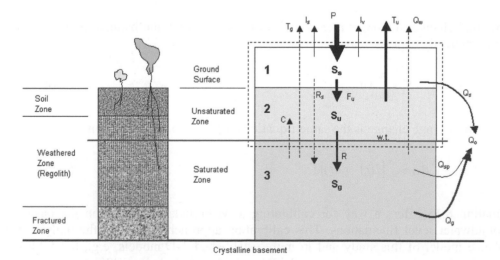

Fig. 6.2. A geological scheme and associated water balance scheme for a crystalline basement aquifer.

Fig. 6.2 shows a conceptual scheme for the catchment daily water balance model. The left part represents the geological scheme indicating the three major geological zones; the soil, regolith and fractured rock. The right hand indicates the corresponding water balance zones; surface, unsaturated and saturated zones.

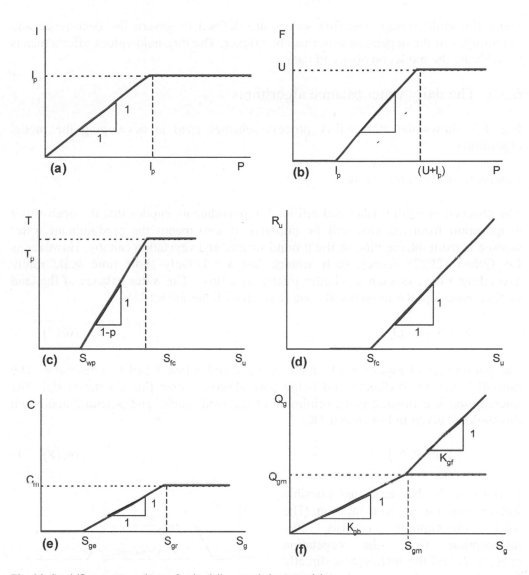

Fig. 6.3. Stock/flux process schemes for the daily water balance model.

Flow is predominantly vertically downwards till the saturated zone. Other flow directions are determined by the nature of partitioning at a partition point. Three partitioning points 1,2 and 3 can be visualized, one in each water balance zone. (See also Section 3.1.2) The first point is in the surface zone. At this point interception, infiltration and surface runoff are defined. Transpiration and recharge (percolation) are defined at the second point in the unsaturated zone. A third partitioning occurs in the saturated zone at which capillary rise, fast groundwater discharge (spring flow) and slow groundwater discharge (baseflow) are defined. The vertical downward output of a water balance zone is the input for the zone below it.

Some basic assumptions are made. The flow in the unsaturated zone is vertical, i.e., there is no interflow or through flow (Ward & Robinson, 1990). Fluxes that occur early in the hydrological process take precedence over the ones occurring later. For example, interception which is the first flux in the balance has the first call on rainfall whilst baseflow, which occurs at the end of the water balance has the last. Transpiration works as a 'pump' on the unsaturated zone and its deficits can be met through borrowings from the saturated zone if set conditions on groundwater storage are met.

Daily threshold storage and flux values are defined to govern the occurrence and magnitudes of the dependent water balance fluxes. The threshold values offer a means to calibrate the model on observed data.

6.3.2 The daily water balance algorithms

Fig. 6.3 shows the storage/flux process schemes used in developing the model algorithms.

Equations of the surface zone

The absence of natural lakes and artificial impoundments implies that the open water evaporation from the area will be minimal. It also means the predominant water storage is from interception on the ground surface and vegetation canopy. However as De Groen (2002) argues, such storage has a relatively short time scale rarely exceeding a day, as such it is better treated as a flux. The water balance of the land surface zone into the unsaturated zone is given by Equation 6.17.

$$F = P - I - Q_s \qquad (6.17)$$

The parameters of Equation 6.17 are as defined in Section 3.2 of this discourse. The rainfall is measured directly and is the only observed input flux for the model. The interception is estimated as the minimum of received rainfall and potential maximum threshold as given in Equation 6.18.

$$I = \min(P, I_p) \qquad (6.18)$$

I_p (mm/d), is the maximum possible interception for a given location. The total interception consists the interception from the vegetation canopy, I_v and the interception directly from the ground surface, I_s. Fig. 6.4 gives the hypothetical variation of the two types of interception through the hydrological year. Interception from vegetation is dominant during the wet season when the vegetation is most thriving. By contrast the soil interception potential is highest during

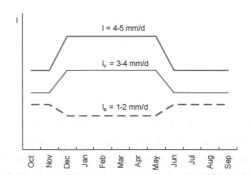

Fig. 6.4. The interception threshold and its components in a hydrological year.

the dry season when the ground surface is driest. The dry and hot conditions in this period imply that low rainfall amounts will not generate enough soil moisture to link with deeper soil moisture and will be evaporated back into the atmosphere. Total interception is given by Equation 6.19.

$$I = I_v + I_s \qquad (6.19)$$

Because of the complimentary nature of the two components of interception as explained above a constant threshold interception can be defined for the whole hydrological year.

The net rainfall, P_{net}, is either zero or the positive difference between the total rainfall and the interception as given in Equation 6.20.

$$P_{net} = \max\left(0, P - I_p\right)$$ (6.20)

In this case the effective rainfall can be defined as the amount of rainfall used for productive purposes either as "green water" in bio-mass production or as "blue water" in surface channels or groundwater storage.

A runoff threshold is defined as the minimum amount of rainfall below which surface runoff is negligible. This runoff is accounted for by considering the infiltration into the unsaturated zone as having a maximum threshold, U.

$$F = \min(P_{net}, U)$$ (6.21)

The infiltration is the input to both the unsaturated and saturated zone water balances. Two possibilities arise for flow to the saturated zone, uniformly distributed percolation through the soil matrix or fast preferential flow through weak zones within the soil matrix. For preferential flow two thresholds can be defined. The infiltrating water must be above a threshold value, U_s, for preferential flow to occur whilst the preferential paths have a maximum capacity, R_{sm}, to transmit the water through the soil matrix. The amount of recharge through preferential flow can therefore be given by Equation 6.22.

$$R_s = \max\left(0, \min(R_{sm}, P_{net} - U_s)\right)$$ (6.22)

where R_s [LT^{-1}] is the preferential flow recharge to the saturated zone.

The surface runoff is calculated as the residual of the surface zone water balance.

$$Q_s = P_{net} - F$$ (6.23)

The runoff generation completes the processes at the first separation point.

Equations of the unsaturated zone

The input fluxes to the unsaturated zone are the infiltration, F, and the capillary rise, C, from the groundwater storage. The outflow fluxes are the recharge to the saturated zone, R, and the transpiration, T_u, through vegetation. The change in storage in the unsaturated zone defines the processes at the second partitioning point and is given by Equation 6.24.

$$\frac{dS_u}{dt} = F + C - T_u - R$$ (6.24)

All the water balance components are as defined in Section 3.2 of this discourse.

The capillary rise depends on the groundwater storage. It can be assumed to vary linearly between zero and a maximum rate determined by the soil type. The capillary rise can be used productively in the root zone if the water table is within an extinction depth, D_{ue} (mm). The extinction depth is the depth below the root zone at which the capillary rise ceases to contribute to transpiration. The capillary rise is maximum at the root depth and zero below the extinction depth, otherwise it is determined from Equation 6.25.

$$C = \min\left(\frac{S_g - S_{ge}}{S_{gr} - S_{ge}}, 1 \right) \cdot C_m \tag{6.25}$$

S_g, S_{ge} and S_{gr} (mm) are respectively; the groundwater storages as a function of time, the groundwater storage at extinction depth and at the root depth. C_m is the maximum capillary rise.

To better account for the influence of vegetation in the water balance of the unsaturated zone this zone is subdivided into the root zone and the percolation zone. In the root zone soil and vegetation characteristics determine the water balance behavior. The soil determines the storage behavior whilst the vegetation determines the rooting depth and the transpiration rate. The soil moisture, S_u, (mm) is the product of the volumetric water content of the root zone, θ_u, (-) and the root depth, D_{ur}, (mm).

$$S_u = D_{ur} \cdot \theta_u \tag{6.26}$$

The root depth can be defined as the depth beyond which native vegetation cannot abstract water from the soil matrix. Below the root zone percolation is the dominant flux. Plants cannot abstract all the water from the root zone. The available moisture, S_{um}, (mm), is that part of soil moisture from which a plant can meet its transpiration demand. Two moisture content values demarcate the available moisture. The field capacity, θ_{fc}, (-), is the maximum amount of water that a freely draining soil holds whilst the wilting point, θ_{wp}, (-), is the water content at which most native plants wilt irreversibly (Schroeder, 1984).

$$S_{um} = D_{ur} \cdot \left(\theta_{fc} - \theta_{wp} \right) \tag{6.27}$$

For this study the field capacity and wilting point for a soil were taken as the moisture content values of the soil when its pF values from a standard pF[23] curve are 2 and 4.2 respectively (De Laat, 2001). For silty sands these pF values translate to soil moisture contents of 0.21 and 0.02 respectively.

The plant transpiration is assumed to be zero below the wilting point and equal to its maximum potential above the field capacity. In between these two limiting values transpiration is assumed to vary linearly with soil moisture as given in Equation 6.28.

---

[23] pF is the negative logarithm of the soil suction pressure.

$$T_u = \min\left(\frac{S_u}{(1-p)\cdot S_{um}}, 1\right)\cdot \alpha_u \cdot T_p \qquad (6.28)$$

The coefficient, α_u (-), accounts for the proportion of the land covered with shallow rooted vegetation when a catchment is treated as a lumped parameter. The fraction, p is the proportion of soil moisture not readily available to the plant (Doorenbos & Kasan, 1979). A p value of 0.5 is often assumed in literature (Savenije, 1997).

For diffuse vertical flow through the unsaturated zone groundwater recharge occurs only when soil moisture is above field capacity and its magnitude equals the excess soil moisture as given in Equation 6.29.

$$R = \max\left(0, S_u - S_{fc}\right) \qquad (6.29)$$

The soil moisture in a given time step is the sum of the soil moisture from a previous time step and the sum of the fluxes over the time step and is given by Equation 6.30.

$$S_u(t) = S_u(t-1) + (F + C - T_u - R)\cdot \Delta t \qquad (6.30)$$

Equations of the saturated zone

At the third partition point the change in groundwater storage in a time step is given by Equation 6.31.

$$\frac{dS_g}{dt} = R - C - T_g - Q_w - Q_{sp} - Q_g \qquad (6.31)$$

All fluxes are in mm/d and as defined in Section 3.2. Two outflow components of Equation 6.31, direct abstraction from the groundwater storage by deep-rooted vegetation, T_g, and human abstractions, Q_w, do not depend on the magnitude of groundwater storage for their occurrence and magnitude. Only the rooting and well depths limit the occurrence and magnitude of T_g and Q_w.

The transpiration by deep-rooted vegetation that tap from the groundwater is estimated by Equation 6.32.

$$T_g = \alpha_g \cdot \mu_g \cdot T_p \qquad (6.32)$$

α_g is the proportion of land covered by deep-rooted vegetation, μ_g is the contribution of the groundwater to transpiration by deep rooted vegetation and T_p is the potential transpiration by the deep rooted vegetation.

The human abstractions can be approximated by Equation 6.33.

$$Q_w = h_p \cdot N \qquad (6.33)$$

N represents the population size and h_p (mm/d) the daily per capita consumption.

The dominant fluxes from the saturated zone storage are the fast groundwater discharge through springs, Q_{gf} and the slow groundwater discharge directly feeding the river, Q_{gb}. The occurrence and magnitude of these two discharge fluxes from the groundwater storage depend on the storage in the saturated zone. A maximum groundwater storage threshold, S_{gm}, is the groundwater storage above which spring flow starts. The spring flow, Q_{gf}, has a shorter time scale, K_{gf}, and acts as a fast groundwater flow component. The spring flow is given by Equation 6.34.

$$Q_{gf} = \left(\frac{S_g - S_{gm}}{K_{gf}} \right) \qquad (6.34)$$

The baseflow is linearly related to the groundwater storage and has a lower limit of zero The maximum groundwater storage is given in Equation 6.35.

$$Q_g = \max \left(\frac{S_g}{K_{gb}}, 0 \right) \qquad (6.35)$$

where, K_{gb} (d) is the time scale of the slow groundwater flow component.

The minimum groundwater storage is fixed at zero and defined as the groundwater storage at which baseflow ceases. However, well and deep-rooted vegetation abstractions can continue to abstract thereby creating negative groundwater storage conditions. The model thus permits groundwater mining.

The storage at a given time step is the sum of the fluxes over the time step and the storage from the previous time step.

$$S_g(t) = S_g(t-1) + (R - C - T_g - Q_w - Q_{gf} - Q_{gb}) \cdot \Delta t \qquad (6.36)$$

The reservoir system

The above formulations can be represented as a system of reservoirs. The shallow nature of the groundwater and the small thickness of the aquifer mean a simple system can be developed. The time scale of processes is relatively short and responses almost immediate. This reduces the need for intermediate transfer reservoirs. For example for deep groundwater tables several reservoirs may be required to simulate percolation and correctly represent the delayed response in the water table fluctuations. Fig. 6.5 illustrates the simple reservoir system for the water balance model discussed here.

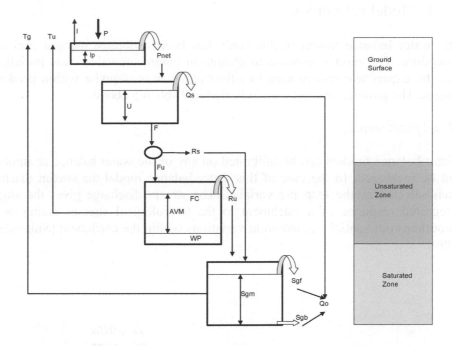

Fig. 6.5. System of reservoirs for the water balance model of a small crystalline basement aquifer.

Assumptions of the water balance calculation

Model calibration yielded a set of parameter values upon which the water balance of this section is based. Table 6.5 summarizes these values.

Table 6.5. Physical properties and process thresholds.

Water balance zone	Parameter	Symbol	Unit	Calibration value	Min. value	Max. value
Unsaturated zone	Root depth (D_r)	D_r	m	2.5	2.0	--
	Field capacity	S_{fc}	-	0.21	0.15	0.3
	Wilting point	S_w	-	0.02	0.0	0.1
	Transpiration factor	a	-	0.5	0.3	0.7
	Maximum preferential flow	R_s	mm/d	5	0	30
Saturated zone	Specific yield	S_y	-	0.21	0.18	0.25
	Fast gw* discharge coefficient	K_{gf}	d	5	3	10
	Slow gw discharge coefficient	K_g	d	120	100	150
	gw separation point	S_{gm}	mm	10	8	15
Process thresholds	Interception threshold	D	mm/d	5.0	3	6
	Infiltration threshold	U	mm/d	70	50	--
	Capillary rise	C	mm/d	1.0	0.5	2

* = groundwater

6.3.3 Model calibration

The water balance model of this study has been calibrated using a trial and error procedure. The model response to change in parameter values was initially studied and the experience gained used to adjust parameters stepwise within predetermined ranges. The parameter ranges are indicated in Table 6.5 above.

Model performance

Water balance models can be calibrated on any of the water balance components that can be measured. In the case of this water balance model the stream discharge has been selected as the response variable. The stream discharge gives the single most integrated response of a catchment to the hydrological stresses acting on it as it smoothens out spatial and temporal variations within the catchment (Silberstein et al., 1999).

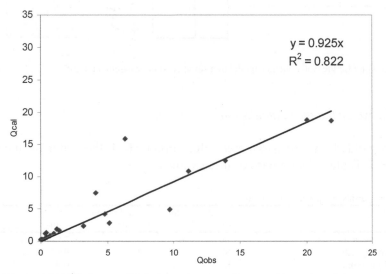

Fig. 6.6. Model fit for monthly discharge (mm/month) over a 48 month period.

Fig. 6.6 shows the fit for monthly observed stream discharge data and model estimates over four years. A coefficient of determination, R^2, of 0.8023 has been obtained implying a 90% correlation between observed and modeled estimates.

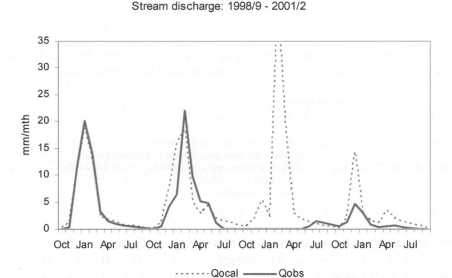

Fig. 6.7. Monthly time series for observed and modeled stream discharge.

Fig. 6.7. shows the observed and modeled monthly time series. (It should be noted that there were no observations in 2000 due to failing equipment). For a prescribed error criterion of ±15% for the water balance estimate the model performed satisfactorily. The deviation of model results from the observed values can be ascribed to the stage method used to measure discharge which can miss floods above the capacity of the gauge. Such floods are captured in the model. It follows that the model results may give a fairer estimate of the true discharge compared to the observed values.

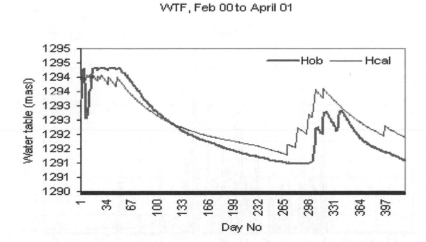

Fig. 6.8. Daily time series for observed and modeled groundwater level fluctuations.

Groundwater levels offer a secondary check on the water balance model. A plot of the accumulated changes in groundwater levels gives an indication of the movement of the water table over a hydrological year. Unlike the discharge however, groundwater levels tend to be localized. The performance criterion can therefore be relaxed compared to the discharge. Fig. 6.8 shows that the model can simulate groundwater levels.

Model response and sensitivity

Table 6.6 summarizes the parameters that describe fully the water balance model. The parameters have been grouped according to the method of determination.

Table 6.6. Water balance model parameters.

Model parameter	Method of determination
S_{fc}, S_{wp}, C, p, C_m,	Established practices from literature
S_y	Based on literature corroborated by field observations
I_p, U, T_p, R_{sm}, U_s, D_r, K_{gf}, K_{gb}	Calibration parameters constrained by expert knowledge
P, Q	Input/output state variables
S_u, S_g	Model state variables

An observation from the model is that introducing a day delay in the model better simulates peaks in the observed hydrograph suggesting that stream discharge peaks are generated from quick groundwater discharge rather than immediate runoff during a rainfall event.

Data from two rainfall stations have been used in calibrating the model. Chikara is centrally located in the catchment whilst Makwavarara is in the headwater part of the catchment.

The rainfall station nearer to the stream discharge station gives a better fit than the more distant station or the weighted average of both. Second, as Fig. 6.9 shows, some peaks recorded in the observed hydrograph did not correspond with observed rainfall. This confirms that storms are of small spatial extent such that the rainfall gauge may miss runoff-generating storms within the catchment. The temporal and spatial variability of the rainfall is therefore important in modeling the water balance of such a catchment.

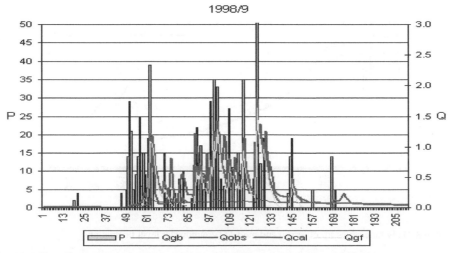

Fig. 6.9. Effect of temporal and spatial variations on model results (mm/d) at a daily time step.

Interception determines the net rainfall infiltrates into the soil and/or becomes runoff. As such it influences the peak discharges in the model. Interception values below 3.5 mm/d result in overestimation of the stream discharge by 12% and 88%. Above 6

92

mm/d the model underestimates stream discharge. An interception ranges between 4.5 mm/d and 5.5 mm/d gives a good model fit.

The model shows that the generally accepted values of 1.5 to 3 mm/d (Pitman, 1973), based on rainfall throughfall, quoted for interception in the region underestimate interception in catchments like the Nyundo. As this catchment is fairly representative of most catchments in Zimbabwe at least, it suffices to say effective rainfall may be seriously exaggerated (Savenije, 2004).

Values of potential transpiration below 1.6 mm/d result in discharge estimates above 30% of the monthly average. Values of the potential transpiration above 9 mm/d result in a 10% underestimation of discharge. On average the model overestimates discharge by 1% for every 2% reduction in the potential transpiration value.

The root depth defines the soil moisture capacity and hence the transpiration totals over time as well as potential recharge in the case of diffuse flow. A threshold depth of 2 m is obtained from the model. Depths less than 2 m result in an overestimate of discharge and a loss in hydrograph trend. For root depths greater than 2 m discharge reduces by between 2% and 14%.

The field capacity and wilting point have minimal impact on the discharge totals and hydrograph trend. They could be confined to a range between 0.14 and 0.23. A field capacity of 0.21 fitted the model. A value of 0.02 has been used for the wilting point.

The model is sensitive to the saturation zone parameters. The groundwater discharge separation storage, S_{gm}, has a huge impact on the modeled discharge over a small range. A similar impact is shown by percentage of infiltration contributing to preferential flow. A 25% increase in the preferential fraction will cause a 25% increase in the discharge estimate.

Discharge coefficients, K_{gf} and K_{gb} affect the fit of the hydrograph. The obtained values of 5 days and 3 months respectively serve to show that the groundwater residence time in the catchment is relatively short.

Discussion on model calibration and sensitivity

A simple lumped parameter model can describe the water balance of such a catchment sufficiently well even with limited data. However, at a daily time step, spatial and temporal variability in the rainfall can influence the model performance greatly. Some peaks in the observed discharge cannot be fully accounted for by the observed rainfall at some stations. From two stations data, it was observed that the data of the station closer to the weir simulated the observed discharge better than the station further upstream. The closer station is within a radius of 15 km from the gauging weir and the second rainfall station is just outside the 25 km radius from the discharge-gauging weir. In some instances observed discharges could not be explained by observed rainfall even at the closer station.

This phenomenon of missed and unexplained discharges is probably because of the localized nature of the rainfall in the catchment. As explained in the areal reduction factors in Section 5.2 it is possible to have a storm reduce in depth and intensity to 50% of its storm center value within a radius of 12 km. Thus a storm that causes a measurable discharge at the weir may not be captured at the rainfall station used to simulate the daily discharge. A second explanation has to do with the high infiltration capacity of the catchment soils and subsequent low runoff. If rainfall falls away from the gauging weir and infiltrates its impact on the stream discharge will take longer to be observed.

Two important points are highlighted by this observation. To correctly simulate discharge, even in a small catchment like the one here, daily models will require an extensive network of rainfall gauging stations. With a storm areal reduction gradient of say 20% per every 5 km the Nyundo catchment would require about 7 rainfall gauges to capture at least 80% of each storm depth and adequately account for the discharge at a single weir. Given the budgetary constraints of most developing countries such a luxury is not affordable nor is it useful for water resources management purposes. It may even be questioned if accuracy is actually improved by such a dense network.

Interception plays an important role in the water balance of small tropical catchments. The generally accepted low values of between 1 and 3 mm/d exaggerate the amount of discharge generated in the catchments. A more appropriate value should lie between 4 and 6 mm/d. An explanation for this increased interception can be found in the nature of the rainfall in Nyundo and similar areas. Rain storms are of short duration and are often accompanied by high temperatures. This combination of high energy and short storms means intercepted rainfall is quickly evaporated thereby creating more interception storage. A secondary explanation is that most methods for estimating interception focused on the so called canopy interception and completely ignored the contribution of the bare ground.

Soil moisture storage, as defined by the root depth, plays an important role in the determination of transpiration but may not be crucial in the determination of groundwater recharge since the recharge is also through preferential paths.

Uncertainty of model parameters remains an issue in this modeling approach. Needless to say the multiplicity of parameters means that several combinations of parameter values can give a similar result. To limit the model degrees of freedom the approach of this study has been to constrain the parameter values in line with field observations and expert knowledge. (See Table 6.6).

6.3.4 Model output - The Nyundo catchment water balance

The Nyundo annual catchment water balance

This section presents and discusses the annual magnitudes and proportions of the catchment water balance with respect to gross annual rainfall, i.e., the rainfall as measured at a gauge station in a hydrological year. (See also Section 5.2.) Fig. 6.10 summarizes the annual catchment water balance.

Evaporation from interception accounts for up to a third of gross rainfall. This value is higher than generally quoted values for relatively higher rainfall areas. For example, Roberts & Harding (1996) report estimates of 9% and 15% of annual rainfall for bamboo and pine canopy respectively for high rainfall catchments of Kimakia and Makiama in Kenya. However, the high estimate fully accounts for the "loss" term often quoted in literature for semi arid regions and is in agreement with the work of other authors in the region. Smit & Rethman (2000), reported initial loses of rainfall between 40 and 60% from plot scale measurements of indigenous tree covers at selected locations within the semi-arid areas of the SADC region. Though these authors attributed the high losses to runoff, De Groen (2002) correctly argues that these losses can only be interception since the runoff coefficients in the region rarely exceed 10%. In fact during the wet months Interception can be as high as 50% of *E*.

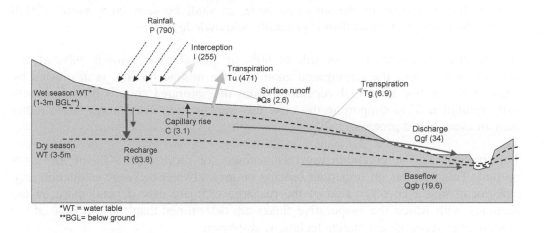

Fig. 6.10. Nyundo average annual catchment water balance (mm/a), 1998-2002.

The correlation between interception and rainfall is about 95% over the modeled period. The obtained results show an almost direct linear relationship between gross rainfall ($^{24}k = 0.391$, $R^2 = 0.9091$) whilst the results from quoted literature suggest that lower rainfall is associated with high proportions of interception whilst high rainfall is associated with lower annual interception proportions. A simple explanation for this behaviour is that interception directly from the soil surface, mulch layer and small pools has an upper limit such that when this upper limit is reached interception accounts for no more losses in the rainfall. Also the low values of interception often quoted for vegetated areas ignore interception direct from the soil surface and therefore still put interception on the lower side of the actual value.

Groundwater recharge estimation approaches need to consider interception. Interception has to be satisfied before all other processes implying that it is most likely to consume its threshold value regardless of the rainfall received (Savenije, 2004). The consumed amount of rainfall is lost to recharge. As such recharge can only be a proportion of net rainfall. Thus the error introduced in expressing recharge as a fraction of gross rainfall is higher in low rainfall areas and periods than in higher rainfall ones. For example, quoting recharge as 10% of annual rainfall results in an error of 40% in the recharge estimate if an annual interception threshold of 200 mm/a is assumed for a 700 mm/a annual rainfall. An argument can be proposed here that since recharge is linked to interception, which in turn depends on the occurrence of rainfall, then recharge can be treated as a stochastic process like the rainfall.

---

^{24} k is the proportionality constant and R^2 is the regression constant.

The importance of interception in water resources management is often underplayed. In this model the interception threshold is higher than the maximum daily transpiration. This may mean that a whole day of biomass production is lost from the first rains after a dry spell as transpiration can only occur after interception. Coupled with preferential flow to the saturated zone, as shall be seen later, more rainfall escapes biomass production than is generally acknowledged.

Transpiration accounts for upwards of 60% of annual gross rainfall. 98% of this amount comes from the unsaturated zone whilst a minute fraction is abstracted by deep rooted and/or perennial vegetation from the saturated zone. Correlation with daily rainfall is 33% supporting the view that transpiration is a storage based rather than an event based process.

Transpiration and interception together account for more than 90% of the gross rainfall. This fact has an important bearing on groundwater recharge estimation as it may mean that a correct value of the recharge estimate may depend more on the accuracy with which the evaporative fluxes are determined than the precision of the groundwater recharge estimation technique employed.

Catchment stream discharge accounts for less than 10% of the gross rainfall. What is important to note however is that the discharge is best modeled by assigning three components to it rather than the traditional two components of baseflow and surface runoff. In addition to the surface runoff and baseflow components a third component from the groundwater is added. Thus, the groundwater discharge is divided into fast and slow flow components. A discharge coefficient of 5 days has been assigned for the fast component and a value of 120 days for the slow groundwater discharge. A fast discharge threshold of 10 mm has been considered. Results show that surface runoff is not significant in the catchment whilst the fast flow component accounts for up to 60% of the stream discharge and the slow component for less than 35%.

The observed stream discharge characteristic presents a new dimension to the recharge debate. The common practice in baseflow separation techniques is to assume that the groundwater discharge gives the minimum estimate for groundwater recharge and recharge is generally defined as the water that replenishes the groundwater. However, the two components of groundwater discharge have different time scales: should they be lumped together as recharge? If so then the available resource for the year is grossly over estimated as 60% of this recharge is only available in the wet season shortly after the rainfall.

Net groundwater recharge has been estimated around 8% of the gross rainfall. 57% of this amount contributes to the fast component of stream discharge, 29% to the slow component and the remaining 14% to direct abstraction by deep-rooted vegetation.

Storage changes at an annual time step are negligible. This point is crucial for the management of water resources in such catchments in general in that there is no natural carry over of groundwater resources from one year to another. Thus the whole concept of long-term recharge estimates is somewhat superfluous in shallow aquifers underlain by hard rock. As shall be evident later understanding the intra-annual dynamics of the water balance is far more relevant than deriving lump sum annual estimates as the former approach better answers the question of when in the year the resource is available.

The negligible storage changes from year to year also imply that baseflow separation may be an ideal method for estimating recharge in such catchments. The baseflow estimate integrates all processes in the catchment and cancels out the small scale temporal and spatial variations within the catchment.

Inter annual variations give an indication of the available resource over time at an annual time step. Large variations imply that resource availability is unreliable. On the other hand, small variations may indicate robustness of relationships from year to year. A measure of variability is the coefficient of variation as defined in Section 5.2. Table 6.7 gives an indication of the coefficients of variation for water balance components as observed in the catchment over the four years modeled.

Table 6.7. Inter annual variations for water balance components in the Nyundo catchment.

Year	P	I	T	R	Q	Q_s	Q_{gf}	Q_{gb}	$\Delta S_u/\Delta t$	$\Delta S_g/\Delta t$
98/99	743.5	242.5	441.6	60.1	50.8	0.0	33.6	17.2	9.3	-1.1
99/00	961.9	352.4	463.0	73.1	60.4	0.0	37.7	22.7	84.1	1.7
00/01	945.3	282.0	530.7	78.3	78.1	10.4	46.9	20.8	55.2	-1.1
01/02	507.3	143.3	475.9	43.7	34.8	0.0	17.0	17.8	-147.5	0.5
Avg	789.5	255.1	477.8	63.8	56	2.6	33.8	19.6	0.3	0.0
%P	100	32	61	8	7	0.3	4.3	2.5	--	--
CV(%)	27	34	8	24	--	200	37	13	--	--

Fluxes that depend more directly on rainfall occurrence tend to have CVs that are higher. Interception, which occurs only when rainfall occurs, and the fast groundwater discharge component with a time scale of about five days after the occurrence of rainfall, have CVs close to that of rainfall. The two fluxes have CVs between 30 and 40% compared to about 25% for rainfall. The surface runoff, which depends on extreme values of rainfall, has the highest CV of around 200%.

The similarities in the CVs of interception and the fast groundwater discharge is because they occur at the same partition point. Recharge is also from preferential flow from the surface. Fluxes that occur at the same partition point can be incorporated into the same hydrological function. Where preferential flow is not overriding then the variations are likely to be different and hydrological relationships can be independent of either flux. The high CV of surface runoff means it can be excluded from hydrological relationships without greatly compromising the validity of results.

Fluxes that depend on storage have lower CVs. This is so because transpiration depends on storage and as such can be sustained over periods with little or no rainfall following adequate rains. Transpiration has a CV less than 10% whilst a value less than 15% has been observed for the stream discharge. In developing model functions for such fluxes the impact of storage cannot be ignored.

It is important to note that the different annual variations for components of the same flux imply different behaviour for these different components. If it is accepted that in modeling a flux different expressions have to be developed for its components the correctness of lumping flux components becomes questionable. For example, "evapotranspiration" combines open water evaporation, bare soil evaporation, interception and transpiration. These processes have different variability and time scales and thus may be better estimated by different hydrological expressions (Savenije, 2004). From the discussion above it makes better hydrological sense to treat interception and fast groundwater discharge separately from transpiration and slow groundwater discharge to correctly simulate the hydrological behaviour of the catchment.

Notwithstanding the arguments on variability as presented above simple relationships for the different fluxes with respect to rainfall can be derived from a regression analysis of the annual values over the four years. The following relationships only serve to illustrate that 'rough' estimates of annual values can be derived from annual rainfall data but do not in any way imply linearity on the hydrological system. The short period of observation also diminish the validity of the expressions as empirical 'rules of thump' but they are retained here as rough guides to annual flux estimates for the catchment.

$$I = 0.4 \cdot (P - 137) \tag{6.37}$$

$$E = 0.45 \cdot (P + 842) \tag{6.38}$$

$$Q = 0.07 \cdot (P - 61) \tag{6.39}$$

The regression coefficients for this linear fit are 0.90, 0.96 and 0.81 for the interception, evaporation (=interception + transpiration) and stream discharge respectively. Transpiration alone exhibits poor correlation with rainfall thus a linear estimate is not possible.

As has been seen from the water balance description above the total groundwater recharge is equal to the groundwater discharge thus Equation 6.41 gives an indication of the annual recharge in the catchment. However, this recharge may be split into the fast discharge and slow discharge components with the fast component being estimated from Equation 6.39. The slow component is the difference between Equations 6.39 and 6.40.

$$Q_{gf} = 0.055 \cdot (P - 175) \tag{6.40}$$

However, as hydrological processes are not linear, minima/maxima functions can be fitted to describe the hydrological regime where the minimum and maximum values denote the threshold values for the described process as explained in the development of the model. The linear Equations, 6.37 to 6.39 can be written as minima/maxima functions as given below.

$$I = \min\{P, 0.4 \cdot (P - 137)\} \tag{6.41}$$

$$E = \min\{P, 0.45 \cdot (P + 842)\} \tag{6.42}$$

$$Q = \min\{(P - E), 0.07 \cdot (P - 61)\} \tag{6.43}$$

Fig. 6.11 is a graphical presentation of the rainfall split in the Nyundo catchment showing the dominance of the evaporative fluxes T and I.

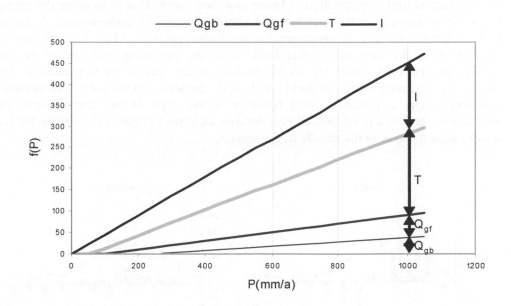

Fig. 6.11. A graphical split of the rainfall in the Nyundo catchment.

The Nyundo seasonal and monthly water balances

Table 6.8 gives an indication of the monthly water balances that have been derived from the daily water balance model.

Table 6.8. The average monthly water balances for the Nyundo catchment, 1998/99 to 2001/2002.

Mth	P	I	T	R	Qs	Q_{gf}	Q_{gb}	$\Delta S_u/\Delta t$	$\Delta S_g/\Delta t$
Oct	21.5	7.8	35.6	1.6	0.0	0.0	0.3	-22.9	0.7
Nov	127.7	43.7	29.9	10.1	0.0	0.6	1.2	45.0	7.4
Dec	158.3	58.0	41.9	12.0	0.0	6.8	2.8	47.8	0.9
Jan	121.3	36.2	51.4	10.2	0.0	7.3	2.6	24.9	-1.1
Feb	226.7	55.4	42.8	20.2	2.6	13.4	2.7	107.7	2.0
Mar	70.5	31.0	68.1	4.7	0.0	4.3	2.4	-32.0	-3.3
Apr	49.0	16.6	62.5	3.9	0.0	1.0	2.1	-33.3	0.0
May	14.5	6.5	39.7	1.0	0.0	0.4	1.9	-32.2	-1.8
Jun	0.0	0.0	25.6	0.0	0.0	0.0	1.4	-25.4	-1.6
Jul	0.0	0.0	27.3	0.0	0.0	0.0	1.0	-27.0	-1.3
Aug	0.0	0.0	20.7	0.0	0.0	0.0	0.7	-20.5	-1.0
Sep	0.0	0.0	32.2	0.0	0.0	0.0	0.5	-31.8	-0.9
Sum	789.5	255.1	477.8	63.8	2.6	33.8	19.6	0.3	0.0
%P	100	32	61	8	0	4	2	0	0
Avg	65.8	21.3	39.8	5.3	0.2	2.8	1.6	0.0	0.0

The hydrological year has a distinct wet and dry season as is characteristic of semi-arid catchments in the tropics. 90% of rainfall occurs in the wet season. Interception behaves in a similar manner. Baseflow and transpiration are the only processes that occur significantly in the dry season. Even so 70% of transpiration is in the wet season and only 28% of the baseflow is observed in the dry season. Such dynamics only highlight the fact that the catchment is dominated by deciduous vegetation and/or dryland agriculture with marginal groundwater utilization for productive purposes.

Fig. 6.12 summarizes the annual time series for the water balance in the Nyundo catchment over the four-year period. The evaporative fluxes, interception and transpiration, dominate the wet season and appear compensatory in that when interception is high transpiration is lower and vice versa. This is so since the energy required for interception cannot be used simultaneously for transpiration. A natural queue exists with the transpiration having lower priority. Thus for low rainfall amounts near the interception threshold, as at the beginning and end of the wet season, interception dominates. In the middle of the season, when rainfall is well above the interception threshold and soil moisture significant, transpiration dominates. The evaporative total however is the sum of the transpiration and interception and in a good rainfall year follows an almost ellipsoidal form for the wet season with its peak in the middle of the season.

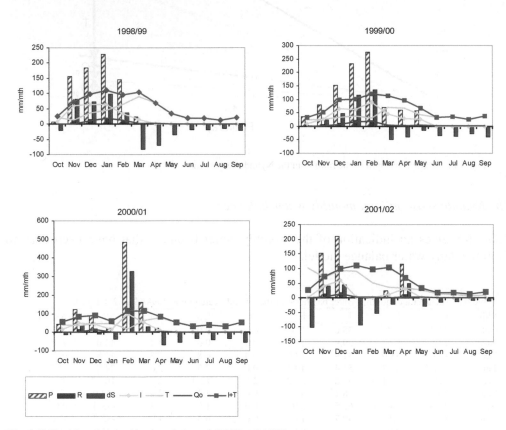

Fig. 6.12 The Nyundo monthly water balance, 1998/9 to 2001/2.

Of note is that evaporation is lower than rainfall in the first half of the wet season but higher in the second half. The significance of this observation is that when modeling transpiration as a function of rainfall at monthly time step soil storage may require greater weight in the later half of the wet season than in the first half. In the first half the memory of the system, as reflected in storage changes, may have a lesser bearing on the amount of transpiration than in the second half. Conversely, groundwater recharge by diffuse flow is most likely to occur at the end of the first half, when sufficient storage has built up in the unsaturated zone, than in the second half of the wet season. In the case of preferential flow it may be that paths become more pronounced as the season progresses.

100

Latent in this simple fact is the idea that if recharge is estimated at a monthly rather than yearly time step the same procedures applied in temperate regions may be applicable to semi arid areas since the argument that potential evaporation is higher than rainfall may not be overriding at this smaller time step. For example, in the month of February the average rainfall is 226 mm/month, the potential evaporation is 4.3 mm/d yielding an initial recharge estimate of 105 mm/month. As De Groen (2001) shows, reasonable monthly estimates of transpiration and interception can be obtained by using the Markov property of rainfall and the number of rain days in a month.

The years 2000/1 and 2001/2 illustrate the impact of rainfall distribution on transpiration. The behaviour of the evaporative fluxes depends on which part of the season the significant rains fall. In the year 2000/1 heavy rains occurred in February after some sketchy falls in the first half of the season. Transpiration was reduced significantly by then. By contrast the year 2001/2 started with good rains followed by a prolonged dry spell. Transpiration did not reduce drastically. As a result the annual totals for transpiration are comparable for the two years even though 2000/1 recorded almost twice the rainfall of 2001/2.

This behaviour is important in agricultural practices for communal societies as it puts them in a dilemma as to when to plant. Early planting with low rains followed by a dry spell can be catastrophic for the communal farmer. Water resources management strategies that seek to capture the early rainfall for use in the succeeding dry spell can thus revolutionize agricultural practices. The simplest, cheapest of such approaches is the enhancement of infiltration to boost groundwater recharge.

Groundwater recharge at a monthly time step is not directly proportional to rainfall nor is it inversely proportional to evaporation. Recharge appears to depend on system memory and is a function of accumulated rainfall. An interesting point from the model is that groundwater recharge in the catchment is better simulated as a combination of preferential flow and diffuse flow through the unsaturated zone. Several issues surface from this behaviour. Preferential flow is often ascribed to tree roots and therefore expected in forested areas. Given that the modeled catchment is mainly Savanna and agricultural, this simple model suggests that other causes for preferential flow such as ant holes, soil cracks, grass roots and the interface between rocky high ground and flat loose soil terrain may actually play a bigger role than usually acknowledged. Surface ponding in cultivated furrow ridges and hoof marks in pastureland could also be contributory factors to the preferential recharge.

A second issue derives from the observation on preferential flow and relates to the estimates of groundwater recharge itself. If preferential paths are dominant then the chloride content of groundwater should be as close as possible to that of rainfall yielding generous estimates of recharge. The water table fluctuation method on the other hand may record an abnormal peak at short time steps before the establishment of equilibrium in the aquifer such that a generous estimate of recharge is again obtained. Agreement of the two methods therefore may not imply correctness of the recharge estimate.

Fig. 6.13 shows that there is a wide variation in monthly processes from year to year. The figure gives the coefficient of variation for monthly rainfall over the four years of observations compared to the long record observations from the Upper Mupfure catchment. The figure serves to show that monthly relationships cannot be restricted to the calendar month but should be based on the rainfall

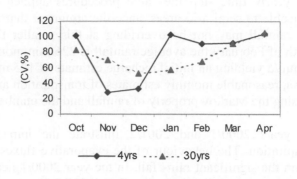

Fig. 6.13. Inter annual variation per calendar month (CV,%).

received. Though the study period is short, and therefore does not quite fit the long record observations, it has the strength that it depicts the likely rainfall scenarios over time in the catchment, i.e., slightly above normal rainfall (1998/9 and 1999/2000), a wet year (2000/1) and a dry year (2001/2). As such the observed variation is probably the widest possible within such a short observation period.

The variations for other water balance components are shown to follow the same pattern as the rainfall.

The Nyundo daily water balances

The water balance at annual and monthly time steps is sufficiently large to mask some temporal and spatial variations for processes occuring with time scales less than a month. The daily balance reveals more of the dynamics and relationships between hydrological components.

Table 6.9. The average daily water balances for the Nyundo catchment (mm/d), 1998/99 to 2001/2002.

WB component	P	I	Q_s	T	R	Q_{gf}	Q_{gb}	$\Delta S_u/\Delta t$	$\Delta S_g/\Delta t$
Maximum	85.4	5.00	10.4	3.20	8.40	2.50	0.20	61.6	6.90
Average	2.16	0.70	0.01	1.31	0.17	0.09	0.05	0.00	0.00

Table 6.9 shows the average and maximum values for the main hydrological processes in the catchment. Wide departures from the mean are the norm reflecting the sporadic nature of processes. Because of this wide daily variation in processes, the estimation of parameters is better done using maximum and minimum values rather than average values. However, this is not to say average values are totally useless as rough estimates of parameter values. For example, a fairly reasonable estimate of daily recharge can be obtained from an average water balance of the unsaturated zone. Subtraction of the transpiration from the net rainfall gives a recharge estimate of 0.17 mm/d. However the drawback still remains that actual evaporation is difficult to determine. But if this could be overcome, the uncertainty in the recharge estimates as shown in the traditional methods could be reduced. The argument here is that it may be beter to invest in methods that give clearer assessments of evaporation than those that seek to measure directly the actual recharge rate.

6.4   Summary and conclusions

The long record rainfall and flow data for the upper Mupfure shows that it is feasible to develop linear relationships for rainfall, discharge and evaporation. Such relationships, coupled with threshold process limits, can be used to develop a daily water balance model for small catchments. Such a daily water balance highlights some important points regarding water resources management in these catchments. At the selected runoff threshold, surface runoff is of little consequence in the catchments. Interception is crucial at low rainfall amounts. Given a mode rainfall of 10 to 15 mm/d the modeled interception threshold implies 30% to 50% of recorded rainfall is lost to interception. Transpiration is a more sustained flux over a daily time step compared to interception and recharge. The later two fluxes have higher maximum values but lower average values. Both recharge and interception are shown to be event based. Thus to enhance recharge may require physical interventions that directly impact on event processes. Maintaining storage however requires greater consideration of vegetation cover.

CHAPTER SEVEN

7.   RECHARGE ESTIMATES IN THE NYUNDO CATCHMENT

This chapter outlines the methodologies applied to estimate the groundwater recharge in the study. It explains the underlying considerations and mathematical expressions used to analyze data for the water balance (WB), water table fluctuation (WTF), and the chloride mass balance (CMB) methods for estimating groundwater recharge.

The methods used in this study were selected because of their wide use in the SADC region. No attempt was made to assess methods that may be 'state of the art' but are not used in the region.

7.1   Recharge from the catchment water balance

7.1.1 Background

The water balance (WB) method generally estimates recharge as the residual of all other components of the water balance (Simmers, 1997) and can be applied equally to the unsaturated zone or the saturated. However, most studies have been carried out in the unsaturated zone primarily to confine the water balance to vertical fluxes and eliminate the effects of lateral saturated zone flows.

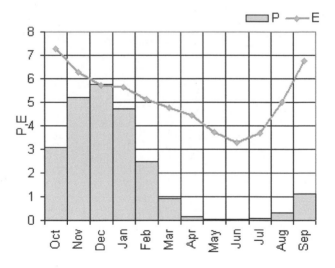

Fig. 7.1. Variation of 20-year average daily rainfall and potential evaporation for the Nyundo catchment (mm/d).

Several authors (Simmers, 1997; Lerner et al., 1990) reckon that the method is best suited for regions where annual rainfall is higher than potential evaporation. Fig. 7.1 highlights the limitation of the method when applied to the study area. The long term daily average rainfall only equals the potential evaporation in December but is generally lower than the latter throughout the hydrological year. This however does not mean the method is not applicable to the study area. Due to the low maximum threshold for evaporation (<5 mm/d) and the large variations in actual daily rainfall it

104

is possible to estimate potential recharge over a short period by taking the water balance at small time steps and then accumulate the values to determine the total for a longer period. A tried approach has been to use ten-day rainfall in a direct water balance (Hendrickx & Walker, 1997). However of late some authors have argued that recharge estimates are greatly improved if a daily time step is used in the water balance.

The main advantage of the WB method is that it does not depend on the mechanism of recharge and that recharge can be estimated from components such as rainfall and stream discharge that are readily measured and are part of most hydrological monitoring regimes.

The main drawback is that the recharge, being smaller than the components from which it is being estimated, is sensitive to the errors in the measurement of those components. For example, Howard & Lloyd (1978) in a study of a temperate chalk aquifer showed that an over-estimate of the potential evaporation by 20% results in an underestimation of recharge by 5%. The sensitivity of the recharge estimate to evaporation is expected to be even higher for semi-arid arid tropical areas. Combined with baseflow separation (BFS) techniques the method integrates the spatial and temporal variations in catchment processes to give an areal average rather than a point value of recharge.

7.1.2 Water balance recharge results for the Nyundo catchment

This section analyses the groundwater recharge in the Nyundo catchment as determined from the catchment water balance presented in Section 6.3.

Occurrence of recharge

The water balance model of Section 6.3 reveals that the dominant mechanism for recharge is preferential flow. This means recharge occurs with the occurrence of rainfall and does not require soil moisture to reach its maximum holding capacity before occurring. This is not to discount diffuse flow recharge, it remains important. It may be that preferential flow is probably more dominant in the earlier part of the season when shrinkage cracks, animal burrows play a more significant part. With progressive wetting such paths become either clogged by the deposition of fine grained soil particles in percolating water or the soil matrix reaches near saturation conditions thus reducing drastically the volume of voids in the matrix.

A conclusion from the water balance is that close to, if not more than, 80% of the recharge is due to preferential flow. Recharge has, under this condition, a direct relation with net rainfall and is less dependent on soil moisture conditions as is the case when recharge is mostly through diffuse flow. The ensuing discussion is based on the conclusion that recharge is dominated and therefore controlled by preferential flow.

At an annual time step the mechanism of recharge is not relevant as the concern is with total values. As the time step becomes smaller the mechanism of recharge becomes important as it determines the delay between the occurrence of rainfall and the response of the groundwater storage. Preferential flow results in short delays mainly due to the degree of wetting required before a crack can convey water. The delay in preferential flow recharge is therefore determined by the depth to the water table. Diffuse flow has a longer delay since the flow depends on hydrological processes in the unsaturated zone (available storage, demand for transpiration, etc.) in addition to the depth to the water table. The time scale for preferential flow can be as short as a day or less whilst that for diffuse flow can be a month or more.

The partitioning of infiltration into preferential flow and unsaturated zone infiltration occurs just below the surface whilst that for diffuse flow occurs in the unsaturated zone. Such a partitioning protocol means that preferential flow is an event-based process and diffuse flow is a storage driven process. The recharge in the catchment can therefore be expressed legitimately as a function of causative rainfall.

At a monthly time step rainfall-recharge relationships become more difficult to determine since the observed recharge has a higher contribution from diffuse flow which is influenced by soil moisture storage. The soil storage dynamics become important. Thus, where diffuse and preferential flows contribute to groundwater recharge the total recharge is given by the combination Equation 7.1.

$$R = R_s + R_u \qquad\qquad (7.1)$$

where, R [LT^{-1}] is the total recharge, R_s [LT^{-1}] is the preferential flow recharge and R_u [LT^{-1}] is the diffuse flow recharge.

In the following discussion however, no distinction is made between the two components of recharge.

Magnitude of recharge

The daily water balance model suggests that the annual gross groundwater recharge estimate is about 8% of the gross annual rainfall. However, it does not seem to follow that a year of high rainfall results in high recharge. Fig. 7.2 shows that years with similar annual rainfall totals can have different values of the annual recharge. The rainfall for the year 99/00 exceeds that of the year 00/01 by about 2 % but the estimated recharge for 99/00 is less by about 1 % compared to the year 00/01. It may be that antecedent years play a role in the recharge for the year. If relatively high soil storage is retained from a preceding year the likelihood of diffuse flow recharge increases in the current year thereby increasing the total recharge.

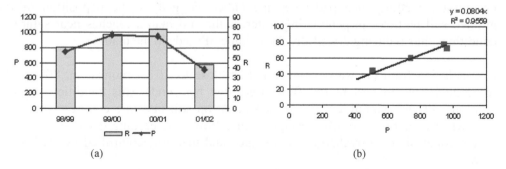

Fig. 7.2. Annual recharge estimates and rainfall (mm/a), (a) totals and (b) relationship.

At a monthly time step recharge, as estimated from the water balance model, is positively related to rainfall. The amount of recharge varies between 4 and 9 % of monthly rainfall. Fig. 7.3 shows the fit between monthly rainfall and monthly recharge.

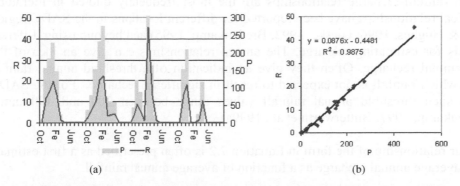

(a)   (b)

Fig. 7.3. Monthly recharge estimate and monthly rainfall (mm/month), (a) temporal variation and (b) linear relationship.

At a daily time step recharge is also correlated to rainfall. Of greater importance however is the frequency distribution of the recharge as it gives an idea of the mean daily recharge and its distribution. The frequency distribution of the rainfall is similar to that of the rainfall. Fig. 7.4 shows that up to 45% of daily recharge is below 1 mm/d and that the daily recharge rate hardly exceeds 10 mm/d.

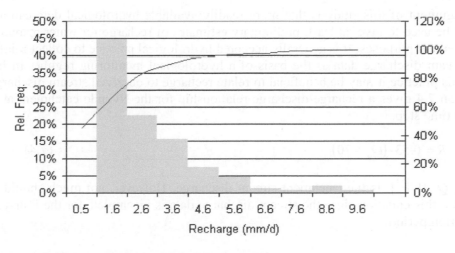

Fig. 7.4. The frequency distribution of daily recharge (mm/d).

Rainfall-Recharge relationships

The water balance model simulates the catchment hydrology but does not offer a directly predictive formula for estimating the recharge. The central question in groundwater resources management is the estimation of groundwater recharge – can recharge be predicted from rainfall at a specified time step? In answering this question it is important to develop some models that can predict the recharge at different time steps.

Annual rainfall-recharge relationships are the most frequently quoted in literature. Empirical relationships have been reported for different locations in the SADC region (Sami & Hughes, 1996; Gieske, 1992; Bredenkamp, 1995) and beyond using different methods for estimating recharge. The annual relationships can give an idea of the gross annual recharge. Often they give an indication of a threshold annual rainfall below which rainfall is not expected to result in significant recharge. For the SADC region such threshold annual rainfall values range between 200 and 400 mm/a (Bredenkamp, 1997; Butterworth et al., 1999a).

A linear relationship of the form in Equation 7.2 is often prescribed as a first estimate of the average annual recharge as a function of average annual rainfall.

$$R = k_1 \cdot (P - k_2) \qquad (7.2)$$

where k_1 [-] is a constant of proportionality and k_2 [LT^{-1}], is the threshold annual rainfall, P [LT^{-1}], is the annual rainfall and R [LT^{-1}] is the annual recharge.

Such an approach applied to the four-year study data yields Equation 7.3.

$$R = 0.12 \cdot (P - 270) \qquad (7.3)$$

Equation 7.3 simply means that the recharge is about 12% of net annual rainfall since the threshold value of 270 mm/a is close to the average annual interception of 255 mm/a obtained from the daily water balance model.

The argument of this study is that more readily available hydrological data can and should be used to give, at least, preliminary estimates of recharge for water resources management purposes. It is generally accepted hydrological practice to collect rainfall and stream discharge data as the basis of a hydrological monitoring regime. In line with this practice it may be beneficial to relate recharge to observed stream discharge. Equation 7.4 gives a recharge-discharge relationship for the Nyundo catchment at an annual time step.

$$R = 0.53 \cdot (Q_o + 70) \qquad (7.4)$$

where, Q_o [LT^{-1}], is the catchment stream discharge. However, not much should be read into this equation since the stream discharge data is incomplete for the four-year simulation period.

7.2   Recharge from the chloride mass balance method

7.2.1 Background

The chloride mass balance method determines recharge by calculating the ratio of chloride concentration in precipitation to that in groundwater (Gieske, 1992).

Several underlying assumptions are outlined for the method to be applicable. (1) The chloride is a conservative tracer, i.e., its concentration is neither diminished nor increased through chemical reactions in the soil, vegetative uptake or increased evaporative action among other things. (2) There are no other sources of chloride to the groundwater other than precipitation. (3) The losses through leaching are minimal, i.e., there is no incremental change in chloride concentration in the vertical direction.

The chloride in rainfall originates from the sea and terrestrial dust and is often termed 'cyclic salt' (Hendrickx & Walker, 1997). The chloride tends to decrease with the distance from the sea. Previously this was attributed to the lessening influence of the sea but it has been realized and accepted that the decrease is due to rainout and moisture recycling (Savenije, 1995).

The method has been used successfully in a range of geological environments including coastal sands (Sharma & Hughes, 1985), colloidal clay and cemented sandstones (Edmunds et al., 1988), crystalline basement terrain (Jarawaza, 1999; McCartney, 1998) and Kalahari sands (Gieske, 1992). However with respect to climate, the method has generally been limited to semi-arid and arid areas where the impact of runoff does not have to be taken into account. Thus, (1) the need to determine infiltration rates separately is eliminated and (2) vertical flow can be assumed.

The method applies equally to the unsaturated zone (Edmunds et al., 1988) and below the water table (Gieske, 1992) so long as the unsaturated/saturated soil systems are considered to have attained steady state or quasi steady state conditions. The method is easy to apply and relatively cheap.

The equation for calculating recharge from the chloride mass balance method can be derived by two approaches, one based on the vertical solute transport and the other on the mass balance of the solute in the soil medium.

The first approach is suited mainly for diffuse flow. The equation for one-dimensional vertical flow of a non-absorbing solute in a non-dispersive and relatively inert system as provided by Bresler (1981) can be applied:

$$J = -\mu\theta \cdot \frac{\partial c_g}{\partial z} + c_g \cdot R + D_s \qquad (7.5)$$

where, J [$MT^{-1}L^{-2}$] is the one dimensional vertical flux, μ [-] is the diffusion coefficient of the flux, θ [-] is the volumetric water content, c_g [ML^{-3}] is the chloride concentration, z [L] is the depth measured from the ground surface, R is the vertical flux of water [LT^{-1}], the recharge, and D_s [$MT^{-1}L^{-2}$] is the chloride dry deposition. Assuming quasi steady state conditions and negligible dry deposition reduces Equation 7.5 to:

$$R = \frac{1}{c_g} \cdot \left(J + \mu \cdot \theta \cdot \frac{\partial c_g}{\partial z} \right) \qquad (7.6)$$

For practical purposes a further assumption is made that diffusion is negligible such that Equation 7.6 reduces to:

$$R = \frac{J}{c_g} \qquad (7.7)$$

The vertical chloride load, J, comes from infiltrating rainfall, P, with a chloride concentration c_p with the assumptions that runoff is negligible and the rainfall is the only source of chloride as stated before. The vertical chloride load is given by:

$$J = P \cdot c_p \qquad (7.8)$$

109

Combining Equations 7.7 and 7.8 yields the CMB recharge estimation equation stated earlier.

$$R = P \cdot \frac{c_p}{c_g} \tag{7.9}$$

The terms are as described in Section 3.2. The mass balance approach is discussed at length in the same section.

Atmospheric deposition of chloride

Atmospheric deposition of chloride refers to the process by which chloride is deposited on the earth's surface from the atmosphere. The natural chloride in the atmosphere originates from the sea and follows the hydrological cycle. Wave action in the sea leads to water vapor and chloride ions being released into the atmosphere and concentrated through cloud formation. Winds blow the clouds overland transporting them over considerable distances from the oceans. Rain forms over land and falls bringing the chloride with it to the ground in raindrops. The amount of chloride in the rainfall is greatly influenced by the type of rainfall, the distance from the sea, wind speed and direction and the roughness of the terrain.

Natural atmospheric deposition is likely to decrease with distance from the sea even if the recorded rainfall remains the same due to moisture recycling over the land surface. Savenije (1995) presents models for moisture recycling and its impact on the atmospheric chloride deposition. The models were originally developed for the Sahel region of West Africa. Rainfall originating from the sea reduces over land according to Equation 7.10.

$$P = P_o \exp\left(\frac{-x}{\lambda}\right) \tag{7.10}$$

where P (mm/a), is the rainfall at a location a distance x (km) from the sea, P_o (mm/a) is the original rainfall at the coast and λ (km) is the length scale of the moisture recycling over land, i.e., the distance over which moisture is used only once. The author proposes a λ value of 400 km for the Sahel. The atmospheric deposition follows similar behaviour:

$$c = c_o \exp\left(-\frac{x}{\lambda}\left(\frac{1-\alpha}{\alpha}\right)\right) \tag{7.11}$$

Where λ and x are as defined above whilst c and c_o are the atmospheric deposition over land and at sea respectively. α is the atmospheric loss coefficient, which is close to the runoff coefficient (Q/P). With α in the order of 10%, the reduction rate of the salt concentration, along a moisture trajectory, is 9 times stronger than that of the rainfall. The model of Equation 7.11 may offer a first estimate of the chloride depositions for inland areas for which coastal depositions are known.

This moisture recycling argument however must be treated with caution. Rainfall formation and occurrence is an erratic phenomenon. In areas where rainfall is both cyclonic and convectional the deposition is dependent on the type of rainfall occurring at the time of measurement. Cyclones that travel at high speeds are likely to have high atmospheric depositions further inland than normal convectional rainfall which consist in part of recycled moisture from the land surface

Gehrels (1999) defines four types of chloride depositions with respect to the estimation of groundwater recharge; wet, dry, bulk and total depositions.

The wet deposition, c_{pw}, refers to the chloride dissolved in the clouds and fall to the land surface during precipitation events like rainfall, dew, hail or snow. Wet deposition is a function of the concentrations of substances dissolved in or scavenged from the atmosphere during rainfall and the volume of rainfall at the point of measurement. The measured value is an integration of atmospheric processes that have a large spatial scale compared to the area for which the wet deposition is required. The wet deposition in this case includes not only the chloride from seawater but all chloride acquired in between the sea and the point of measurement.

The dry deposition, D_s, is the fallout of chloride from the atmosphere in the absence of rainfall. The amount of dry deposition is a function of the atmospheric concentration of gases, aerosols and particles and the velocity at which they are deposited (Lynch et al., 2001).

The dry deposition is greatly influenced by the level of development of an area and its surrounding environs. In industrial areas, gases and aerosols that may contain substantial concentrations of chloride pollute the atmosphere and may lead to high levels of dry depositions. Thus, the location of the area for which the dry deposition is required in relation to atmosphere pollution sources is important in deciding whether or not dry deposition should be considered in the recharge estimate. In areas within or downwind of atmospheric pollution sources the dry deposition must be considered.

The capture of dry deposition is dependent on the natural surfaces being considered. Forested areas tend to have higher capture efficiency than uncovered surfaces. Forested areas also have higher interception rates. Since evaporated water contains no chloride, the chloride originally in the rainwater is deposited on the intercepting surface during the evaporation of interception storage. Finally the dry chloride is shaken to the ground as dry deposition. Dry deposition is therefore positively correlated with interception and will increase with increasing vegetation canopy cover.

The annual total chloride mass deposition per unit area is the sum of the wet and dry depositions.

$$P \cdot c_p = P \cdot c_{pw} + D_s \tag{7.12}$$

where, P [L] is the annual rainfall, c_p [ML^{-3}] is the resultant total chloride concentration in rainfall and D_s [ML^{-2}T^{-1}] is the annual dry deposition.

Due to the difficulties associated with its estimation the dry deposition is often taken as a fixed fraction of the wet deposition. The total chloride concentration based on the wet deposition including an allowance for dry deposition can then be formulated as:

$$c_p = c_{pw} \cdot (1 + d) \tag{7.13}$$

where, d [-] is the fixed dry deposition fraction. In the SADC region, researchers have proposed a value of d between 0.05 and 0.1.

The wet and dry depositions as presented above describe the chloride deposition process. Measurements of this process are provided by the bulk deposition. The bulk deposition, c_{pb}, refers to the chemical depositions captured in a field collector vessel such as a rain gauge. As such, the bulk deposition records all the wet deposition during a rainfall event plus that part of the dry deposition on the collector vessel surface prior to the precipitation event.

The bulk deposition estimate depends on the aerodynamic properties of the collector vessel such as its surface roughness which may not be representative of the natural surfaces and will therefore yield dry deposition values that are at best an indication rather than a quantification of the natural dry deposition. In some cases, the construction material of the collector vessel may influence the bulk deposition estimate. In an experiment to measure the suitability of three types of rain gauge to measure bulk deposition Beekman & Sunguro (2002) concluded that a PVC totalizer rain gauge overestimates chloride concentrations in rainfall and ascribe this to partial breakdown of the PVC due to prolonged exposure to sunlight.

The bulk deposition is often taken as an approximation of the total deposition. It has been observed however that vegetation cover, and hence interception, affects this approximation. For example, Postma (1993) suggest that the total deposition can be twice the bulk deposition in forested areas and approximately equal to the bulk deposition in shrub and non-covered areas. An explanation for this observation could be that in forested areas, the gauge height is below the average canopy level and may not capture all the dry deposition given interception effects and aerodynamic gauge properties as explained above. Hence the bulk deposition will be less than the total deposition. In the case of shrub and grass areas, the rain gauge is at the same level or above the average vegetation height and therefore captures more or less the same chloride deposition as the surrounding vegetation. Bulk deposition estimates in such areas are correspondingly close to the total deposition.

However as demonstrated in Section 3.2 interception and dry deposition have a compensatory effect and interception needs not be accounted for where bulk deposition estimates are used as an approximation for total chloride deposition.

7.2.2 The CMB method in the Nyundo catchment

This section summarizes the results of applying the chloride mass balance method (CMB) in the study area for the 2000/01 and 2001/02 seasons. The first season was a wet year receiving over 130% of the long-term average rainfall of 730 mm/a whilst the second was a dry year with less than 65% of the long-term average rainfall. The section discusses the atmospheric chloride deposition, the chloride in precipitation, the chloride content in groundwater and the chloride mass balance recharge estimates for two years. The questions addressed in the section are; what is the magnitude of the recharge estimate from the CMB method? How is this estimate influenced by the chloride deposition, rainfall variability, interception, and vegetation plus geology factors?

Methods: estimation of c_p and c_g

The chloride concentration in rainfall has been estimated from rainfall samples collected in a 100 mm capacity field PVC collector type rain gauge located at the Chikara Clinic rainfall station. The data collection and processing procedures are outlined in section 5.1. The bulk deposition has been considered. It has been assumed that the bulk deposition is equal to the total deposition.

Rainfall samples were collected daily during rainfall periods and stored in sealed plastic containers for onward laboratory chloride analysis. In this way water losses through evaporation were kept to a minimum and the concentration of the chloride in the collected samples could not be altered. This procedure meant that there was no need to consider water volumes in determining the concentration of chloride in rainfall and the recorded rainfall depth was used directly in the estimation of recharge. The total deposition for a series of rainfall samples was estimated as a weighted average:

$$c_p = \frac{\sum_{i=1}^{n} P_i \cdot c_{pi}}{\sum_{i=1}^{n} P_i} \tag{7.14}$$

where, c_p [ML^{-3}] is the average chloride concentration in rainfall, i denotes a rainfall sample and n denotes the total number of samples.

The chloride content in groundwater has been estimated from the analysis of observation well samples. The saturated zone chloride concentration has been used. Groundwater samples were collected as grab samples from selected monitoring wells and boreholes. The average chloride content for the groundwater recharge zone was taken as the harmonic mean for all values in the zone to minimize the impact of outlier values, particularly the larger values.

$$c_g = N \cdot \left(\sum_{i=1}^{N} \frac{1}{c_{gi}} \right)^{-1} \tag{7.15}$$

where c_g [ML^{-3}] is the average chloride concentration in groundwater, c_{gi} [ML^{-3}] is the chloride concentration in groundwater at an observation well, i, and N is the number of observation wells considered.

Results: The chloride content in precipitation

Table 7.1 summarizes the results obtained from the study area in the 2000/1 and 2001/2 seasons.

The weighted average bulk deposition for 2000/01 was 1.06 mg/l and that for 2001/02 was 1.17 mg/l giving an average concentration of chloride in rainfall of 1.10 mg/l over the two years. Eighty and ninety-five percent of the annual rainfall has been used in the analysis for 2000/01 and 2001/02 respectively.

Table 7.1. Chloride bulk deposition from the Chikara rain station (2000/01-2001/2)

Sampling date	Chloride conc. (mg/l)	Total rainfall (mm/sample)	Average daily rainfall, (mm/d)
2000/01			
10 Dec. 2000	1.55	57.0	2.4
3 Jan. 2001	1.42	30.0	2.3
31 Jan. 2001	2.03	12.5	2.8
11 Feb. 2001	1.11	137.5	15.2
19 Feb. 2001	1.16	115.5	13.0
28 Feb. 2001	1.37	230.0	28.6
14 Mar. 2001	0.21	119.5	13.3
25 Mar. 2001	0.10	40.0	7.1
2001/02			
09 Nov. 2001	1.51	44.0	22.8
27 Nov. 2001	1.21	105.0	9.7
16 Dec. 2001	1.87	149.5	12.3
29 Dec. 2001	1.40	63.0	12.5
29 Jan. 2002	0.95	5.0	1.1
21 Mar. 2002	0.72	25.5	4.2
17 Apr. 2002	0.10	120.0	16.6

(a) Sampling dates.

MONTH	Oct	Nov	Dec	Jan	Feb	Mar	Apr
Conc. (mg/l)	--	1.0	1.6	1.5	1.2	0.3	--

(b) Average monthly values during study period.

Bulk chloride depositions are not usually included in hydrological monitoring networks. As a result very few databases exist for comparison of results. For Zimbabwe, Larsen et al. (2001) proposes a value of 0.5 mg/l for western Zimbabwe whilst Beekman & Sunguro (2002), quote values of between 0.28 mg/l for Bulawayo in the 1991/92 season and 0.35 mg/l for Marondera in the 1995/96 season. Other recent studies, notably Sankwe (2001) and Mabenge & Nyamuranga (2001) report values of between 0.8 and 1.2 mg/l for Marondera during the 2000/01 and 2001/02 seasons.

Elsewhere in the SADC region Beekman et al. (1996) suggest values between 0.2 and 1.0 mg/l for Botswana whilst Nkotagu (1996) quotes values between 2.0 and 2.8 mg/l in Tanzania. Sami & Hughes (1996) however suggested that the mean of the large rainfall events associated with recharge was around 2.2 mg/l. Their recommendation has been adopted as an upper limit for rainfall deposition in several subsequent studies. For example, Beekman & Sunguro (2002) argue that the chloride deposition for Zimbabwe cannot exceed 1.0 mg/l despite field observations to the contrary. Edmunds et al. (2002) also reject a value of 1.77 mg/l in favour of 0.65 mg/l in Maiduguri, Northern Nigeria. Both authors suspect analysis procedures for the large values derived from field observations.

The average chloride deposition value of 1.1 mg/l obtained in this study is low in comparison to most areas outside Zimbabwe. As the location is at least 500 km from the sea the low values suggest considerable moisture recycling and rainout. However this value is still high compared to values suggested from similar studies in the country. A reason for this difference could be that the chloride deposition depends on the part of the season when the samples are taken and the percentage of rainfall captured for the analysis. As such, the quoted values above depended greatly on the period when assessments were made and the percentage of rainfall used for the analysis.

It should also be noted that the field value may be influenced by the rain gauge type, sampling and laboratory procedures adopted in determining the deposition (Sankwe, 2001).

Fig. 7.5 shows the observed variations in wet depositions over the two years of recorded data. The chloride bulk deposition shows a diminishing trend during the course of the season. Two reasons may be offered for this trend. The first reason is that the early season rainfall contains dry deposition from the dry season. Second, the rainfall between December and April is partly due to cyclones (Makarau, 1995). Indeed, cyclone Bonita was observed in February 2001. Such cyclones bring rain direct from the Indian Ocean and the chances of high wet deposition are higher compared to the normal conventional rainfall in which repeated moisture recycling plays a larger role in rainfall occurrence.

A decreasing trend as the season progresses is evident for the two seasons. If we accept that these two seasons are representative of the long-term trend then we can conclude that the chloride deposition decreases as the season progresses. Thus it is crucial to note when during the season measurements of wet deposition are taken. Early season measurements may lead to higher estimate of recharge whilst late season measurements will yield a lower estimate for recharge.

A second observation is that the chloride bulk deposition has no discernable correlation to either the rainfall or the seasonal rainfall trend. The relation between rainfall depth and chloride concentration is random. As such recharge estimates can only be made for the whole year at the end of the wet season since high rainfall periods within the season do not correspond to periods of high or low chloride deposition. The recharge estimates at a smaller time step than a year are therefore highly variable and bear no relation to the actual occurrence of recharge.

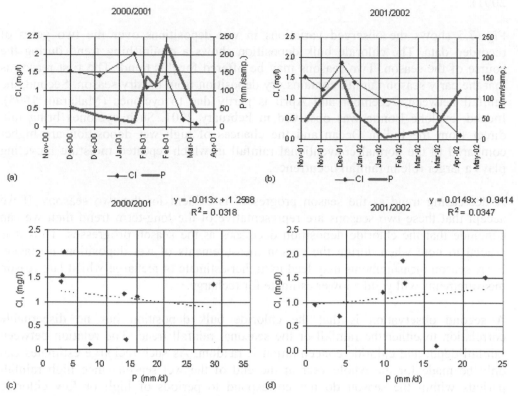

Fig. 7.5. Seasonal variation in chloride bulk deposition from the Chikara rain station (2000/01-2001/2). (a) temporal variation of bulk deposition and sample rainfall depths for 2000/01, (b) temporal variation of bulk deposition and sample rainfall depths for 2001/02, (c) scatter plot for bulk deposition against daily average rainfall over each sampling period in 2000/01 and (d) scatter plot for bulk deposition against daily average rainfall over each sampling period in 2001/02.

A third important, but rather obvious, point is that the chloride deposition is independent of the total rainfall received during the season. The average chloride concentrations from the two years are almost equal despite that the rainfall in the first year is almost double that of the second season. Such an observation may be misleading. Since the origin of rainfall has greater influence on the chloride concentration it is quite possible that a dry year with only cyclonic rainfall due to the IOC system can yield a higher chloride concentration than a wet year with predominantly convectional rainfall due to the ITCZ. Hence recharge estimates from one year's observations may be grossly misleading.

It follows from the above observations and arguments that the long-term average chloride deposition is an important but as yet a rather indeterminate parameter in estimating recharge using the CMB method. In the SADC region, at least, long observation periods are still required before confidence can be generated in the final estimate. Such observation need to measure the concentration of chloride in rainfall over the season and for as much of the total annual rainfall as possible in addition to documenting the origin of the rainfall over the season. The dynamics of rainfall need to be known if the recharge estimate is to be improved.

Given that the rainfall in the region is caused by two systems, the ITCZ and the IOC which as discussed in Section 5.2 dominate at different stages of the season, the estimated wet deposition is greatly influenced by the period from which samples are

116

taken. As a recommendation, not only should as much of the annual rainfall as possible be included in the analysis but the sampling should be spread as much as possible over the season.

The values for the 2000/01 and 2001/02 are, in line of the above arguments, reasonable. The annual average bulk deposition value of 1.10 mg/l has been used in the determination of recharge in this study.

Chloride concentration in groundwater

The second important parameter in estimating groundwater recharge by the CMB method is the chloride concentration in soil and/or groundwater. Two approaches can be adopted, either the soil water of the unsaturated zone is used or the groundwater of the saturated zone is used.

The advantage of the saturated zone concentration is that due to groundwater flow the groundwater sample represents a spatially integrated sample with a smaller time scale since it results from mixing concentrations over a wider area compared to the soil water sample which represents the dynamics within a vertical column over a longer time scale (Grismer et al., 2002; Davisson, 2000). Second, the groundwater samples are easier and cheaper to collect and analyze compared to soil moisture samples. For these reasons groundwater samples were used for determining recharge in the study.

A minimum groundwater chloride concentration of 4.3 mg/l and a maximum of 25.5 mg/l has been obtained with a range of 21.2 mg/l and mean of 9.15 mg/l. The coefficient of variation (CV) has been estimated at 60%. Twenty-four observation wells have been used in the analysis.

Table 7.2 gives a comparison of the Nyundo data with selected locations in granitic basement areas in Zimbabwe. The observed groundwater chloride concentrations, c_g, are in general agreement with records from the national groundwater quality database for nearby and similar areas. The table shows that the groundwater chloride concentration for most granitic catchments has a wide variation suggesting that local factors influence the groundwater chloride concentration. The Nyundo minimum and average values are on the high side of the range when compared to other locations whilst the maximum value is in the middle of the range

Table. 7.2. Some groundwater chloride concentrations in granitic basement aquifers in Zimbabwe (mg/l).

Location	Min	Mean	Max
Buhera	0.5	1.4	31.9
Chikomba	3.9	7.9	15.2
Goromonzi	0.2	1.5	68.2
Harava	2.0	5.7	290.0
Hwedza	1.0	5.8	36.0
Marondera	1.0	3.3	45.0
Murehwa	2.3	5.5	94.2
Nyundo	4.3	9.1	25.5

of the other values. The minimum value suggests chloride accumulation in the Nyundo catchment.

Vegetation has an impact on the concentration of chloride in groundwater, especially where perched water tables and localized flow exist (McCord et al., 1997). However, Böhlke (2002) noted that the major source of chlorides, among other chemicals, in groundwater is agricultural fertilizer particularly potassium chloride (KCl) rather than natural vegetation.

For Nyundo it has been noted that the cultivated areas have a higher concentration of groundwater chloride compared to pasture and forested areas. (See Table 7.4 below).

Such behavior suggests that the impact of fertilizer use on the groundwater chloride concentration in the area cannot be ignored.

The sandy soils in the area encourage fertilizer use and the leaching of its chloride component to the groundwater cannot be ruled out. Lamond & Leikam (2002) recommend that for soil chloride concentrations above 6.0 mg/l no fertilizer need be applied. This limit is well below the values observed for the groundwater chloride concentration in the Nyundo catchment. Chloride leaching from the root zone and subsequent accumulation in the groundwater may explain the high chloride in the groundwater of the area.

The impact of geology on the groundwater chloride concentration cannot be easily distinguished from that of topography in the area. The groundwater chloride concentration is relatively higher in higher altitude southeastern parts of the catchment compared to lower parts of the catchment. The higher parts are associated with basalts and bended ironstones and the lower areas with Karoo sands. (See also Section 5.3). The basalt and BIF complex may have higher chloride concentrations due to lithological make up and also the lower infiltration rate which leads to chloride accumulation.

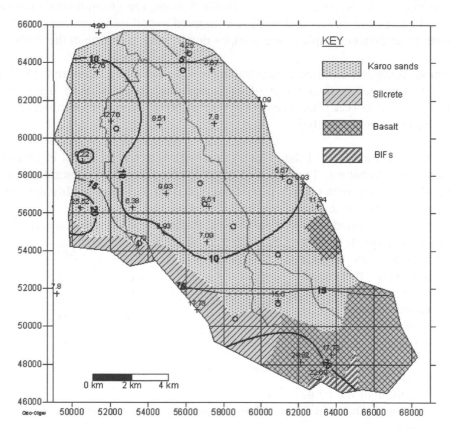

Fig. 7.6. The spatial distribution of mean groundwater chloride concentration (mg/l) in relation to geology in the Nyundo catchment. Values are the average for September 2000 and 2001.

Fig. 7.6 illustrates the groundwater chloride distribution (contour lines) and lithological variation (shading) in the Nyundo catchment.

118

Weathering influences groundwater chloride content. In the upper parts of the catchment the water table is deeper. Above the water table alternate moist and dry conditions occur during the wet season resulting in chemical weathering. In the lower parts of the catchment the groundwater table is shallow resulting in near permanent saturation and therefore little or no chemical weathering.

As can be recalled from the discussion of Section 5.3 the regolith is thinnest in the upper parts of the catchment. These areas exhibit a wider variation in the geology. Compared to areas of more homogenous lithology, i.e., the Karoo sands of the lower parts of the catchment, the chemical weathering is more active in these upper parts. However, as the area is predominantly Karoo sand and sampling was mainly in shallow wells only penetrating these sands, the impact of basalts is minimal on the regional chloride distribution.

In conclusion, it has been shown that, in the case of the Nyundo catchment, the groundwater chloride concentration appears to have a correlation with land use, topography and geology. Particularly, the use of fertilizers may explain the high concentrations of chloride in groundwater. The high concentration in the higher parts and low concentrations of groundwater chloride in the lower parts suggests that localized flow dominates over regional flow in the catchment. The same distribution conforms to land use practices, i.e., cultivation (and fertilizer use) in the upper parts and pasture (natural vegetation) in the lower parts. Chemical weathering may also play a part but is more difficult to justify. The challenge for recharge estimation is therefore, to devise methods that filter out the anthropogenic effects to determine the chloride content due only to rainfall.

The distribution of groundwater chloride suggests higher recharge in the lower parts of the catchment. This suggestion should be taken with caution. The lower part of the catchment is a seepage zone, and because of the vertical upwards groundwater flow the CBM method does not apply. (See also earlier discussion in Section 3.2).

Recharge estimates from the CMB

The groundwater recharge has been calculated from Equation 3.53. A wet deposition of 1.1 mg/l and groundwater chloride concentration of 9.15 mg/l has been used in the calculations implying that the annual recharge estimate will be 12% of annual rainfall.

Table 7.3 summarizes the recharge estimates for the four-year study period. A wide range of 55.4 mm/a (or 59% of the average value) between a minimum of 61.8 mm/a and maximum of 117.2 mm/a has been observed. Applying the chloride concentrations of this study to the long period (30 years) data from Mubayira and Beatrice suggests a long-term annual recharge in the area of about 90 mm/a.

Table 7.3. Annual recharge estimates in the Nyundo catchment from the CMB method (mm/a).

Year	Rainfall	Recharge
1998/99	743.5	90.6
1999/00	961.9	117.2
2000/01	945.3	115.2
2001/02	507.3	61.8
4-year avg.	789	94.6

The recharge estimate is sensitive to the accepted value of the wet deposition. For example, if the accepted long-term chloride deposition of 0.4 mg/l (Beekman & Sunguro, 2002) and the obtained average groundwater chloride concentration of 9.2 mg/l are used a recharge of 32 mm/a (or 4% of annual rainfall) is obtained. This simple calculation highlights the sensitivity of the recharge estimate to the assumed value for the wet deposition. Thus in areas where a chloride deposition monitoring network is in its infancy or non-existent recharge estimates by the CMB method are reliant on the assumed value of wet deposition.

Table 7.4. CMB recharge estimates at individual observation wells in the Nyundo catchment.

OW	X	Y	Lithology	Cg (mg/l)	2000/01		2001/02	
					R(mm/a)	R(%P)	R(mm/a)	R(%P)
BH23	50450	56280	Weath. Gr.	25.5	39.3	4	23.3	5
W53	62110	48130	Basalt	24.82	40.4	4	23.9	5
BH52	63080	47230	Basalt	22.69	44.2	5	26.2	5
BH61	57490	60800	Karoo Sa.	7.80	128.5	14	76.1	15
BH20	53460	54300	Weath. Gr.	17.73	56.5	6	33.5	7
W36	56600	50880	Weath. Gr.	17.75	56.6	6	33.5	7
W50	63750	48530	Basalt	17.50	56.3	6	33.5	7
W47	60910	51370	Weath. Gr.	15.6	64.2	7	38.0	8
W43	51290	63490	Karoo Sa.	12.76	78.5	8	46.5	9
W44	52030	60910	Karoo Sa.	12.70	78.4	8	46.5	9
W14	63020	56380	Karoo Sa.	11.34	88.4	9	52.3	10
W16	62200	57560	Karoo Sa.	9.93	100.9	11	59.8	12
W19	54870	54940	Karoo Sa.	9.95	101.0	11	59.8	12
W25	54970	57050	Karoo Sa.	9.90	100.8	11	59.8	12
W45	50530	58780	Karoo Sa.	9.22	108.7	11	64.4	13
W41	54620	60700	Karoo Sa.	8.51	117.7	12	69.7	14
BH30	57220	56780	Karoo Sa.	8.56	117.9	12	69.7	14
BH38	49200	51730	Karoo Sa.	7.80	128.5	14	76.1	15
W37	60170	61690	Karoo Sa.	7.09	141.3	15	83.7	17
W46	57100	54470	Karoo Sa.	7.11	141.5	15	83.7	17
W24	53150	56300	Karoo Sa.	6.38	157.1	17	93.0	18
W60	57370	63650	Karoo Sa.	5.67	176.7	19	104.7	21
W42	51380	65580	Karoo Sa.	4.96	202.0	21	119.7	24
W28	56080	64560	Karoo Sa.	4.25	235.8	25	139.7	28

Table 7.4 shows the recharge estimates at the different observation wells over the two years of observation. The average annual recharge is 12% of the annual rainfall with a minimum of 4% and maximum of 25% of the annual rainfall. The spatial CV is 60%. The high spatial variability implies that spatial averaging from multiple sites is required to obtain a reasonable regional estimate for recharge.

Fig. 7.7 highlights the spatial variability of the recharge based on the average annual estimates for the two seasons. No clear demarcation with landuse, vegetation cover and geology is observable. The northwestern parts of the catchment are more undulating and are dominated by Karoo sands

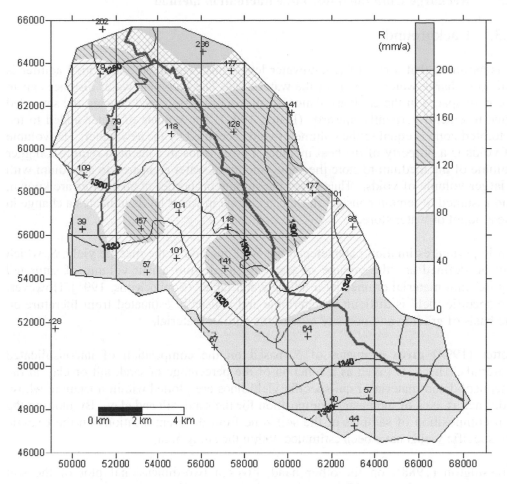

Fig. 7.7. Annual recharge map of the Nyundo catchment for 2000/01 based on the CMB method (mm/a).

Conclusions

The chloride deposition obtained in the study is rather high in comparison with values proposed by other authors for Zimbabwe. However neither has a convincing argument been offered yet to justify the correct chloride deposition in the country nor enough studies carried out to determine the correct deposition. As such the chloride deposition remains the most likely source of the largest error in determining recharge from the CMB method. The groundwater chloride concentration in the Nyundo catchment is somewhat high in comparison with similar crystalline basement terrains probably due to the use of fertilizers. A regional recharge of 12% of annual rainfall has been obtained for the area. The estimates for individual points are highly variable.

7.3   Recharge from the water table fluctuation method

7.3.1 Background

It is assumed that a rise of groundwater levels observed in an unconfined aquifer is due to recharge water arriving at the water table. The incoming water displaces air in the pore spaces in the capillary fringe, displacing the fringe and causing the saturated zone to extend vertically upwards (Fetter, 1994). The volume of water added to the saturated zone is equal to the volume of voids occupied by the new water. The volume of voids is a property of the host medium. Fewer voids in a medium require a bigger volume of the medium to store the same amount of water compared to a medium with a larger volume of voids. Thus when the storage properties of an aquifer are known, and assumed to remain constant in time, a change in water levels indicates a change in the amount of water stored.

An important estimation parameter in the WTF method is the specific yield, S_y which can be defined as "the volume of water released from a unit volume of saturated aquifer unit material drained by a falling water table" (Sophocleous, 1991). However, the specific yield is difficult to determine and is usually estimated from literature on the basis of grading and porosity tests of the aquifer material.

Fetter (1994) gives estimates of S_y based on the composition of unconsolidated materials. The S_y is given as a function of the percentage of sand, silt or clay in the unconsolidated material. Equi-specific yield lines are plotted within a triangle whose sides act as axes of increasing composition for the sand, silt and clay. By plotting the soil composition of samples of the soil zone from different locations on the triangle the specific yields have been estimated within the study area.

Brassington (1988), on the other hand, gives a two-dimensional plot of the soil properties (porosity, specific retention and specific yield) as functions of grain size. With porosity and particle size distribution data as input the specific yield for a given soil sample can be estimated.

In crystalline basement aquifers the S_y is generally considered to be less than 1% and values between 0.001 and 0.02 are often quoted in literature (Chilton et al., 1995; Lovell et al., 1998; Healy & Cook, 2002). The low S_y is attributed to the fractured nature of the basement and the spatial heterogeneity of the aquifer material. Due to the fractured nature, recharge can occur rapidly through the fractures leading to localized high water tables whilst surrounding areas remain relatively unsaturated (Healy & Cook, 2002). Second, the orientation and distribution of fractures may lead to low horizontal permeability and hence prolonged groundwater discharges during the dry season (Price et al., 2000).

The second important parameter is the recession constant, K [T-1] defined here as the residence time for the groundwater system.

Principle of the WTF method

Fig. 7.8 is a graphical representation of the WTF method. The water balance for the saturated zone can be written as explained in Section 3.2 of this discourse.

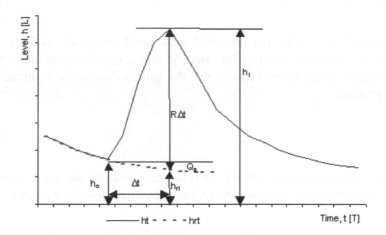

Fig. 7.8. The groundwater level hydrograph for estimating recharge using the WTF method.

$$R = Q_g + T_g + \frac{dS_g}{dt} \pm Q_{lg} \qquad (7.16)$$

As explained above, if the groundwater recharge occurs in a finite time period, dt [T], resulting in a finite rise in the water table, dh [L] and if all other fluxes may be disregarded then the change in storage is given by Equation 7.17 in which S_y [-] is the specific yield of the aquifer. The term dh is set equal to the difference between the peak of the rise and the level when the hydrograph starts to rise.

$$R = \frac{dS_g}{dt} = S_y \cdot \frac{dh}{dt} \qquad (7.17)$$

If the start of the recharge process is set at time zero and a discrete time increment thereafter measured as Δt, the discrete form of Equation 7.17 can be written as:

$$R \cdot \Delta t = S_y \cdot \left(h_t - h_0 \right) \qquad (7.18)$$

where, h_t [L] is the lead of the hydrograph, h_o [L] is the height at the start of the recharge event and Δt [T] is the time step. The total recharge of an event is measured at the peak of the hydrograph (See Fig. 7.8).

In most applications however, the recharge is estimated over a period of days for which the groundwater discharge may be substantial and cannot be neglected. The recharge in this case is the water required to increase the storage as well as account for depletion (Healy & Cook, 2002):

$$R = Q_{net} + \frac{dS_g}{dt} \qquad (7.19)$$

123

where the term Q_{net} [LT^{-1}] is the resultant of the fluxes in Equation 7.16. To account for Q_{net} dh should be larger and taken as the difference between the peak of the rise and the low point on the extrapolated antecedent recession curve. The discrete form of Equation 6.16 in this case is written as:

$$R \cdot \Delta t = S_y \cdot \left(h_t - h_{rt} \right) \qquad (7.20)$$

where h_t [L] and h_{rt} [L] are the observed peak level and level on the extrapolated antecedent recession curve after a time step Δt. Since the recession curve follows an exponential decay under natural conditions, the value of h_{rt} can be estimated from the recession equation:

$$h_{rt} = h_0 \cdot e^{-\frac{\Delta t}{K}} \qquad (7.21)$$

where K [T] is the recession constant (or residence time) for the groundwater system depleting under natural conditions. The recharge can therefore be written as:

$$R \cdot \Delta t = S_y \cdot \left(h_t - h_0 \cdot e^{-\frac{\Delta t}{K}} \right) \qquad (7.22)$$

Note that in the case of direct rainfall recharge the recharge term $R \cdot \Delta t$ can be related to the net rainfall depth.

$$R \cdot \Delta t = \int_0^t (P - I)\,\mathrm{d}t$$
(7.23)

where I [L] represents all rainfall that does not contribute to recharge, mainly interception.

Equation 7.23 can be re-written as a rainfall function by combining it with Equation 7.22.

$$\int_0^t (P - I)\,\mathrm{d}t = S_y \cdot \left(h_t - h_0 \cdot e^{-\frac{\Delta t}{K}} \right) \qquad (7.24)$$

For an aquifer that discharges naturally through streams or springs and for which the discharge can be measured, water level changes and stream discharges can be combined to estimate the recharge over time (Ketchum et al., 2000). The change in level in such a system is due to the difference between the recharge and the net baseflow from the stream and springs (Johansson, 1987).

$$S_y \frac{dh}{dt} = R(t) - Q_{gb}(t) \qquad (7.25)$$

The change in storage with respect to water levels over time can be given by the following integration relationship.

$$S_y \int_{h(0)}^{h(t)} dh = \int R(t)dt - \int Q_{gb}(t)dt \tag{7.26}$$

The left hand term can be represented by a single water level change over the recharge period, T whilst the discharge can be estimated from the groundwater depletion equation.

$$S_y \Delta h(T) = \int R(t)dt - Q_0 \int \exp\left(\frac{-t}{K}\right)dt \tag{7.27}$$

Rearranging Equation 7.27 yields the recharge estimate for a naturally depleting system.

$$\int R(t)dt = S_y \Delta h(T) + Q_0 \left(1 - \exp\left(\frac{-t}{K}\right)\right) \cdot K \tag{7.28}$$

For a time step of unity Equation 7.28 can be further simplified.

$$R_t = Q_{t-1} \cdot K \cdot \left(1 - \exp\left(\frac{-1}{K}\right)\right) + S_y \Delta h_t \tag{7.29}$$

Equation 7.29 can be applied to daily time steps and the total recharge over a period of continuous recharge estimated as the accumulated sum of the daily recharge estimates.

The WTF method is best suited to (1) shallow unconfined aquifers in which vertical water movement is not confined and rises and falls in the water table are sharp and (2) short time periods, in the order of days to weeks, in which recharge is predominant over discharge (Healy & Cook, 2002). Other limitations are that the water table fluctuation may be due to causes other than recharge, such as barometric pressure changes, lateral subsurface flows, groundwater abstractions and wave motion or level changes in surface water bodies (Todd, 1959; Sophocleous, 1991).

The major drawback with this method is its sensitivity to the chosen value of specific yield for the aquifer (Nonner, 2003). The spatial non-homogeneity related to preferential recharge also compromises the representativeness of the estimate from the method over a wide area. However, the attractiveness of the method remains in its simplicity of use, that it integrates the recharge over a local area rather than a point and that it is independent of the mechanisms of recharge in the unsaturated zone (Healy & Cook, 2002).

The method has been used mainly to estimate recharge (Gerhart, 1986; Hall & Risser, 1993) but also for determining the aquifer specific yield (Sophocleous, 1991), estimating total evaporation (White, 1932), and direct evaporation from the ground (Weeks & Sorey, 1973).

7.3.2 The WTF method in the Nyundo catchment

In this section the recharge for the 2000/01 and 2001/02 seasons has been estimated from water table fluctuations data. The recharge has been estimated from two sets of data. Automatic gauge data has been used to estimate the recharge from daily data at two locations, B1 and B2 (See also Section 5.3). Manually recorded data has been used to estimate the recharge at the same boreholes and wells used in the chloride mass balance method to enable a comparison of the two methods at the same location and also account for spatial distribution of recharge in the catchment.

The dynamics of recharge and its relationship with rainfall as determined by the WTF method are investigated at different times steps from the year to the day. The water levels are combined with the water balance model of Section 6.2 to both validate the model and to further investigate the rainfall-recharge relationships.
Methods

Three approaches have been used to estimate the S_y for the various lithologies in the catchment: (1) soil grading and standard soil mechanics tables, (2) literature references for similar environments and (3) analytical calculations based on the water balance and the observed groundwater depletion curves.

Table 7.5. Lithological classification and specific yield in the Nyundo catchment.

Stratigraphic layer	Lithological description	Areal distribution (%)	Mean Specific yield S_y (-)
Soil zone	Karoo sand	60-70	0.015-0.02
	Red clayey silt	10-15	0.01-0.018
	Mixed sand/silcrete	10-15	0.025-0.05
Weathered & fractured zone	Decomposed granite	80-90	0.01-0.022
	Fractured matrix	10-20	0.001-0.002

Table 7.5 shows the general lithologies in the study area, their distribution and the selected values of specific yield. The estimates of the specific yield over the study area show a gradual decrease in S_y from the upper part of the catchment to the lower parts in line with the change in soil texture.

The recession constant, K, has been derived from the semi-logarithmic plots of the recession segments of the groundwater level fluctuation time series. An average recession constant of 100 days has been used in all the calculations.

To estimate the recharge at a given point Equation 7.29 has been applied at a daily time step as well as the event scale for automatically gauged data and at a monthly time step for manually gauged data.

Occurrence and magnitude of recharge

Table 7.6 summarizes the monthly and annual recharge estimates obtained from the daily water table fluctuations[25] for the two automatically gauged observations wells, B1 and B2, for the two years monitored.

Table 7.6. Monthly groundwater recharge in the Nyundo catchment from daily water table changes. (S_y = 0.015)

Month	2000/01 Season					2001/02 Season				
	Rainfall (mm/month)	Rain days	Recharge mm	mm/d	%P	Rainfall (mm/month)	Rain days	Recharge mm	mm/d	%P
Oct	44.0	5	0.0	0.0	0.0	0.0	0.0	0.0	0.0	0.0
Nov	123.0	13	0.0	0.0	0.0	152	14	3.7	0.3	2
Dec	87.4	13	0.0	0.0	0.0	210.3	16	11.8	0.7	6
Jan	22.1	6	0.0	0.0	0.0	3.5	3	5.1	1.7	146
Feb	485.6	26	30.7	1.2	6	0.2	2	0.9	0.4	426
Mar	162.6	15	30.1	2.0	19	24.9	4	1.4	0.4	6
Apr	20.6	5	3.1	0.6	15	116.0	7	1.2	0.2	1
Totals	945.1	83	64	--	6.8	507.3	46	24.1	--	4.8
Avg.	135	12	9.1	0.5		72.5	7	3.4	0.5	

(a) Observation well B1

Month	2000/01 Season					2001/02 Season				
	Rainfall (mm/month)	Rain days	Recharge mm	mm/d	%P	Rainfall (mm/month)	Rain days	Recharge mm	mm/d	%P
Oct	44.0	5	0.0	0.0	0.0	0.0	0.0	0.0	0.0	0.0
Nov	123.0	13	0.0	0.0	0.0	152	14	1.4	0.1	1.0
Dec	87.4	13	0.0	0.0	0.0	210.3	16	31.1	1.9	15.0
Jan	22.1	6	0.0	0.0	0.0	3.5	3	11.2	3.7	320-
Feb	485.6	26	63.7	2.5	6.3	0.2	2	0.0	0.0	0.0
Mar	162.6	15	5.7	0.4	18.5	24.9	4	0.3	0.1	1.0
Apr	20.6	5	0.4	0.1	14.9	116.0	7	0.5	0.0	0.0
Totals	945.1	83	69.9	--	7.4	507.3	46	44.6	--	8.8
Avg.	135	12	10.0	0.4	--	72.5	7	6.4	1.0	--

(b) Observation well B2

The annual recharge for the lower observation well, B2, is estimated at between 44.6 and 69.9 mm/a or 7 to 9% of the total annual rainfall when an S_y of 0.015 is used. The corresponding values for the upper well are 24.1 and 64.0 mm/a or 5 to 6 % of the total annual rainfall. The average recharge monthly rate varies between 3 and 10 mm/month whilst the daily rate is between 0.4 and 1.0 mm/d. the recharge rates are higher in the lower parts (B2) than the higher parts (B1) of the catchment.

It may be that the groundwater levels in the lower parts are influenced by factors other than rainfall recharge. Groundwater flows from the upper to the lower parts and/or

---

[25] Recharge estimates for the early part of the 2000/1 season are missing since recording of groundwater levels commenced in end January 2001. The annual recharge estimate is an under estimate of the true recharge value.

inflows from nearby streams are possible factors. The location of observation wells for determining recharge for the WTF method is therefore important.

The table shows that the monthly recharge is positively but not directly related to the monthly rainfall. The amount of rainfall required to cause recharge in the early part of the season is significantly higher than the rainfall required to induce recharge in the period between December to February. For example, 152 mm of rainfall in November results in only 1.4 mm of recharge in observation well B2 in the 2001/02 season whilst a 38 % increase in rainfall (210.3 mm) in December results in 31.1 mm of recharge, a 20 fold increase in recharge. The delay in recharge suggests a predominance of diffuse flow as a mechanism for recharge.

Table 7.7. Event based gross recharge (mm) from automatic groundwater level loggers (2000/01 season).

End Date	Spell length (d)	Spell rainfall (mm/spell)	Spell recharge					
			B1			B2		
			(mm)	R/P(%)	(mm/d)	(mm)	R/P(%)	(mm/d)
13/2	10	137.0	3.3	2.4	0.3	24.2	17.7	2.4
20/2	7	112.9	9.0	8.0	1.3	16.8	14.9	2.4
26/2	6	153.6	12.3	8.0	2.1	15.1	9.8	2.5
3/3	5	124.6	15.7	12.6	3.1	10.9	8.7	2.2
13/3	6	64.7	10.3	15.9	1.7	0.9	1.4	0.2
23/3	4	37.4	7.7	20.6	1.9	1.2	3.2	0.3
5/4	4	20.3	2.2	10.8	0.5	0.4	2.0	0.1
Totals	42	650.2	60.5	9.3	--	69.5	10.7	--
Average	6	92.9	8.6	9.3	1.6	9.9	10.7	1.4

Table 7.7 shows the results of using the recharge event as a single time step in estimating the recharge for observation wells B1 and B2 for the 2001/02 season when a constant S_y of 0.015 is used. There is general agreement with the daily water table fluctuation approach. It can be observed that the daily rate of recharge is constant for the recharging events during the periods of high recharge. However it does not follow that a high rainfall spell results in high recharge.

The estimates for the recharge can be affected by the technique used to determine the storage changes. For example, the two approaches of this study, the event time step and the daily water table fluctuation yield different results at low recharge occurrences when the hydrograph is not well formed and fitting a recession curve is more difficult.

Table 7.8 gives a comparison of the recharge from the two methods for the same spell periods. With a correlation of 90% the two approaches generally give the same estimate for groundwater recharge. However, in the early part of the recharge period when the recharge peaks are more pronounced the estimates from the event time step method gives an underestimate of recharge whilst in the second part of the recharge period it gives

Table 7.8. Comparison of the hydrograph and daily fluctuation methods for determining recharge over a rainfall spell. (mm)

Date	P	R_{day}	R_{event}	diff.*
02/13/01	137	24.2	21.5	3.6
02/20/01	112.9	16.8	11.8	5.0
02/26/01	153.6	15.1	11.2	3.9
03/03/01	124.6	10.9	14.0	-3.1
03/13/01	64.7	0.9	3.8	-2.8
03/24/01	37.4	1.2	8.8	-7.6
04/05/01	20.3	0.4	2.4	-2.0

an over-estimate of recharge. The main source of error is the value selected for the recession constant, K. The sharper fall in the recession after a recharge event implies that an average recession constant will be less than the true value. The net recession discharge is therefore underestimated. The reverse is the case in low recharge periods.

The WTF method does not only suffer from the handicap of determining the S_y but from the approach selected for calculating the recharge. By introducing a second parameter, K, the estimate error is compounded.

Recharge dynamics

Fig. 7.9 shows the temporal variations in recharge, storage change and rainfall at daily and monthly time steps for the two years of observations for B2.

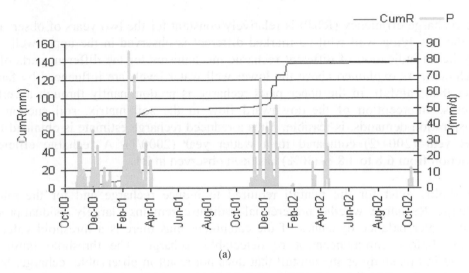

Fig. 7.9. Temporal distribution of recharge and rainfall: daily rainfall and cumulative daily recharge.

The figure suggests that (1) there is a threshold annual rainfall below which water table changes are not pronounced, (2) once the threshold has been overcome the water table response to rainfall becomes evident (3) sharp and short duration rises occur in the water table in response to large rainfall events and (4) the groundwater response to rainfall is delayed.

In the early part of the season, from October to early December recharge occurs but its magnitude is low such that the response of the groundwater levels may not be detectable. Values less than 0.2 mm/d suggesting water level changes less than 10 mm for an S_y of 0.015 have been observed. The daily observations show that recharge can occur as early as November. Since saturated soil conditions will not yet be prevailing, such occurrence suggests the existence of preferential flow as a significant mechanism for groundwater recharge in such a catchment. Preferential flow does not rely on the storage memory of the unsaturated zone.

Observations show that the main recharge events occur in the middle of the rain season, between December and February. In this period moisture conditions influence the occurrence of recharge even at a time step comparatively larger than the time scale of the recharge process itself. This fact is quite apparent for the data of 2001/02. The dry spell in the season from February to March results in reduced recharge in April

129

when significant rains finally fell. Compared to 2000/01 when such a dry spell is not experienced the recharge in April is higher for significantly lower monthly rainfall.

It can be concluded from the above observations that the occurrence of recharge is strongly related to the occurrence of rainfall. However because of the daily time step used in the analysis the relation between recharge and rainfall intensity cannot be evaluated.

The recharge efficiency, defined here as the proportion of gross rainfall that becomes recharge (Ketchum et al., 2002), does not vary as much as the annual rainfall. Thus even though the absolute difference in recharge between a "dry year" (2001/2) and a wet year (2000/1) is about 40 % of the 2000/1 value the change in recharge efficiency remains below 20%. See Table 7.6.

The recharge efficiency (R%P) is relatively constant for the two years of observation for the low lying well whilst a marked different is observed in the upper well. This may be an indication of different recharge mechanisms for the different parts of the catchment. As explained above the lower well water levels are influenced by factors other than rainfall. In the upper parts recharge is predominantly through a vertical process. Interception of the downward flow in the soil matrix, soil suction and transpiration demands, is feasible. Thus a reduced recharge estimate is obtained for a drier year (2001/2) compared to a wetter year (2000/1). A recharge efficiency reduction from 6.8 to 4.8 % (30%) has been observed in B1.

Thresholds exist for the rainfall required to induce recharge and for the rate of recharge. Recharge, whether preferential or diffuse, remains a largely residual process in the water balance dynamics of the catchment thus there is a threshold value for rainfall before commencement of detectable recharge. The threshold rainfall is defined in this study as the rainfall that does not result in observable recharge. In the case of preferential flow the threshold value is due to the wetting required in the soil mass before vertical water movement can occur.

Table 7.9 shows that the annual rainfall threshold varies between 230 and 258 mm/a for the lower well and 258 and 295 mm/a for the upper well. The reduction in absolute value for the threshold between the two years is small, about 13% for both wells, in comparison to the reduction in rainfall of 43%. The threshold is due to processes that have a low inter annual variability probably interception. In fact when compared to the results of the water balance model the threshold average of 260 mm/a is within ± 2 % of the average annual interception of 255 mm/a obtained from water balance calculations.

Table 7.9. Annual groundwater recharge in the Nyundo catchment from daily water table changes (mm/a).

| Year | Annual Rainfall | | |
	Total	Threshold (mm)	%P
B1			
2000/01	945.1	294.8	31.1
2001/02	507.3	257.9	50.8
B2			
2000/01	945.1	258.4	27.3
2001/02	507.3	230.7	45.5

Daily rainfall amounts less than 3 mm/d do not seem to cause any recharge. This observation suggests that there is interception from the soil. The catchment is not heavily forested so canopy interception is low. If soil interception were low much of the rain would recharge the groundwater via preferential flows.

130

These observations suggest that the magnitude of recharge depends on processes that reduce the rainfall before recharge commences. Thus, the importance of considering interception in recharge estimation is highlighted.

Apart from the threshold rainfall, the actual daily recharge rate seems to have an upper limit. An average of 2.4 mm/d has been observed for both observation wells. See Table 7.7. It is a function of the aquifer geological structure and the hydrological process that control the water balance of the unsaturated zone.

The aquifer response to rainfall may explain how the recharge process relates to the aquifer physical properties.

In the recharge period, December to February, the relation between rainfall and recharge occurrence is more pronounced. Nearly all rainfall events induce a response in groundwater levels. However, as Table 7.10 shows, the magnitude of recharge is not directly proportional to that of rainfall nor does the recharge due to a given rainfall event occur only during the day of the rainfall. In the table

Table 7.10. Recharge occurrence during a rainfall spell.

Date	P(mm/d)	R(mm/d)
02/07/01	32.4	0.0
02/08/01	12.5	0.0
02/09/01	12.9	5.5
02/10/01	44.3	6.4
02/11/01	0.1	9.1
02/12/01	1.8	2.2
02/13/01	2.3	0.8

the rainfall of 32.4 mm on the 7th of February does not result in recharge until three days later whilst the rainfall over the four-day spell from the 7th to the 10th continues to recharge the groundwater three days after the end of the rainfall. At a daily time step therefore, the magnitude of rainfall is not related to the amount of rainfall received. Simple rainfall-recharge relationships are therefore better developed at a larger time step to fully account for the larger process time scale.

The magnitude of recharge appears to be determined by the antecedent rainfall and therefore soil moisture conditions prior to the recharging event. In Table 7.6 the 12.9 mm on the 9th appears to cause a recharge of 5.5 mm whilst the 44.3 mm of rainfall on the 10th appear to cause a recharge of 6.4 mm. When moist conditions prevail less rainfall is required for transpiration leaving more rainfall for recharge.

The depth to the water table influences the groundwater response to rainfall. The time lag between the event and observed recharge can be taken as a parameter of the system to rainfall. Fig. 7.10 suggests that the time lag between a rainfall event and the groundwater response decreases with a decrease in the depth to the water table. A linear fit is apparent. Below 3 m the correlation between

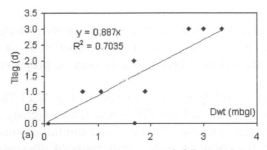

Fig. 7.10. Groundwater response to rainfall events in relation to depth to water table.

the groundwater response and the rainfall event becomes weak. The effect of the rainfall event remains apparent at most 3 days after the event. These observations suggest that diffuse flow cannot be totally discounted in the catchment. As the water table rises the unsaturated zone thickness reduces thus the delay in response by the water table becomes shorter. The longer effect after rainfall occurrence is due to the slow downward movement of the water towards the water table.

The response to rainfall in manually monitored wells corroborated the observations from the automatically gauged data. Positive changes in groundwater levels generally start in January and February. The deeper the depth to the water table the longer the lag.

Rainfall-recharge relationships

From the above observations the response of the groundwater fluctuations has a lag of almost three months from the onset of the rain season. The annual recharge period occurs in the middle half of the season, from the beginning of December to February even though some recharge occurs outside this period.

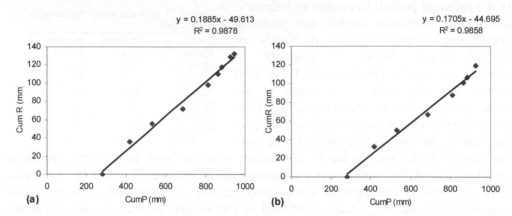

Fig. 7.11. The relationships between rainfall and recharge from the WTF method. (a) cumulative recharge and rainfall for B1and (b) cumulative recharge and rainfall for B2.

Fig. 7.11 illustrates the annual rainfall-recharge relation for the two automatic gauges over the two years of monitoring. A constant annual threshold rainfall can be considered in the catchment such that from Table 7.9 and Fig. 7.11 the gross annual groundwater recharge can be estimated by a linear model of Equation 7.30.

$$R = 0.2 \cdot (P - 280)$$
(7.30)

where, R (mm/a) is the annual gross recharge and P (mm/a) is the gross annual precipitation. Houston (1988) reported a similar finding for crystalline basement aquifers in the Masvingo province. He proposed a threshold value of 400 mm/a.

The cumulative mass curves of Fig. 7.11 (a) and (b) suggest the existence of a threshold annual rainfall below which recharge is not discernable as explained in the arguments above. The threshold rainfall accounts for the rainfall used for interception and transpiration in the early parts of the season before the soil moisture conditions are sufficiently replenished to sustain these fluxes and allow recharge to occur. The threshold rainfall does not discount diffuse flow but merely suggest that the evaporative fluxes have a higher priority for rainfall. The high coefficient of determination, R^2, of 0.99 indicates strong correlation between annual rainfall and annual recharge.

The spatial distribution of recharge

The annual recharge estimates at several locations have been compared to investigate the recharge distribution with respect to land use, topography and geology.

Table 7.11 summarizes the annual recharge estimates at 25 different wells from the WTF method. The recharge varies between 2 and 17 % of annual rainfall with an average of 9 %. The spatial coefficient of variation is 50%. The estimates suggest that the recharge is higher in the lower parts of the catchment than in the upper parts.

OW	X	Y	Z (masl)	$*D_{wt}$ (mbgl)	2000/01 R_{yr} (mm/a)	R/P (%)	R/D (mm/m)	R_m (mm/mth)
BH23	50450	56280	1308.1	2.33	125	13	53.6	41.7
W53	62110	48130	1300.0	2.82	26.4	3	9.4	8.8
BH52	63080	47230	1368.8	2.67	15.6	5	5.8	5.2
BH61	57490	60800	1287.1	2.32	75.9	8	32.7	25.3
BH20	53460	54300	1330.2	3.5	62.2	7	17.8	20.7
W36	56600	50880	1324.9	8.3	110.3	12	13.3	36.8
W50	63750	48530	1338.0	10.67	51.6	5	4.8	17.2
W47	60910	51370	1334.1	9.88	141.6	15	14.3	47.2
W43	51290	63490	1282.4	2.33	65.0	7	27.9	21.7
W44	52030	60910	1282.9	2.84	131.0	14	46.1	43.7
W14	63020	56380	1330.3	5.29	34.2	4	6.5	11.4
W16	62200	57560	1332.9	11.84	89.7	9	7.6	29.9
W19	54870	54940	1326.0	3.3	17.6	2	5.3	5.9
W25	54970	57050	1314.5	2.35	100.0	11	42.6	33.3
W45	50530	58780	1300.0	2.82	60.2	6	21.3	20.1
W41	54620	60700	1286.5	2.58	89.4	9	34.7	29.8
BH30	57220	56780	1300.0	2.87	164.2	17	57.2	54.7
BH38	49200	51730	1308.0	2.09	125	13	59.8	41.7
W46	57100	54470	1340.0	1.91	149.9	16	78.5	50.0
W24	53150	56300	1310.4	1.66	54.3	6	32.7	18.1
W60	57370	63650	1291.8	2.05	100.4	11	49.0	33.5
W42	51380	65580	1279.5	2.61	61.5	7	23.6	20.5
W28	56080	64560	1286.9	2.42	74.7	8	30.9	24.9
**B1	56200	54500	1317.5	3.52	60.5	6	17.2	20.2
**B2	57700	56500	1294.3	2.91	89.9	9	30.9	30.0
Average	---	---						

Table 7.11. WTF recharge estimates at individual observation wells in the Nyundo catchment.
* End of dry season (Sept.)
** Automatic data loggers

Fig. 7.12 shows the spatial distribution in WTF recharge (contour lines) in relation to vegetation cover and land use (shaded areas) in the Nyundo catchment.

No correlation between the recharge estimate and the vegetation cover and, indirectly land use, is observable. This result is rather surprising given that the recharge process is believed and has even been shown to be greatly influenced by land use (Lovell et al., 1998).

Some plot scale studies have shown that recharge is influenced by land use practices that enhance infiltration (Lovell et al., 1998) and by vegetation types that have high transpiration rates.

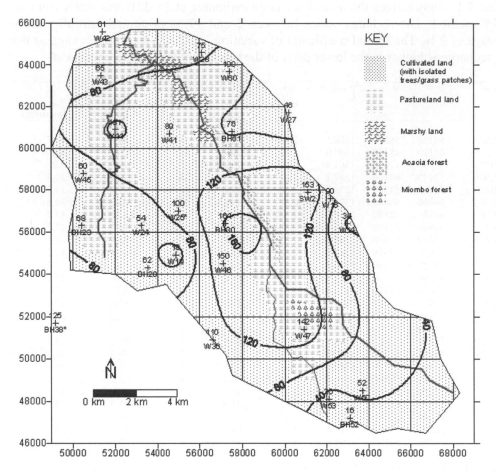

Fig. 7.12. Distribution of recharge from the WTF method in relation to vegetation and landuse in the Nyundo catchment.

The observation in this study suggests that the impact of landuse factors at a larger spatial scale may be limited to the unsaturated zone only and may play a bigger role if recharge was purely through diffuse flow. However, the argument for preferential flow through macro pores due to vegetation roots cannot be fully substantiated since the vegetation distribution in the catchment is not rigid. For example, cultivated land is a mixture of isolated wood patches and cropped land and therefore is not the same as cleared land. Preferential flow may be occurring at the same rate with the same distribution in the catchment.

Fig. 7.13 shows the spatial distribution of recharge (shading) in relation to the topography (contour lines) in the catchment.

Some correlation between the recharge estimate and the topography exists in the catchment. Higher recharge rates of between 100 and 164 mm/a have been observed in the central parts of the catchment. Recharge in the upper and low-lying parts of the catchment is generally low between 34 and 89 mm/a.

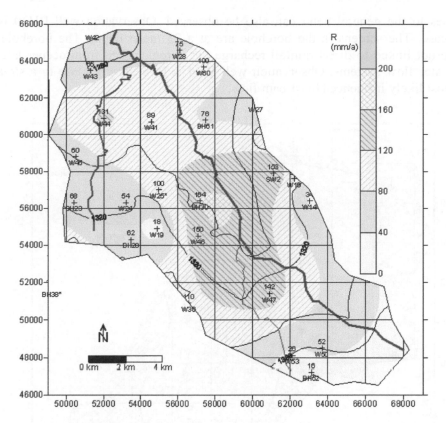

Fig. 7.13. Distribution of recharge from the WTF method in relation to topography in the Nyundo catchment.

A clear pattern is exhibited. High recharge rates occur at the interface of sloping and flat ground. Thus the slope enhances recharge. Lower slopes favor recharge and result in a form of "mountain front recharge" mechanism in which the runoff, most probably interflow, from the interfluves percolates deeper into the aquifer instead of flowing to the surface flow channels.

The generally held view that the recharge zone in a catchment is in the upper parts and the discharge zone in the lower parts whilst the middle parts comprise mainly a transmission zone is not entirely true for such a catchment. It appears from the observations of this study that a better determinant for recharge possibility and potential is slope change rather than mere elevation. A change from higher slope to low slope (concave land configuration) implies higher recharge in the lower altitude areas. Table 7.12 shows some of the slope changes and recharge differences in the Nyundo catchment.

Table 7.12. The relation between slope change and recharge in the Nyundo catchment.

Upper OW	Lower OW	*Diff. In Rch (mm/a)	**Slope change (%)
W19	W25	82	48
W19	W46	132	133
W24	W25	46	50
W36	W46	40	30
W50	W47	90	20
W18	SW2	63	120

*Lower OW – Upper OW
**(Low – high)/low

The difference in recharge rates is not well correlated but it gives an indication of increased recharge in the lower well.

135

Anomalies to the general trend can also be observed. Observation well BH30 is a special case. The screens in the borehole are at a 25 meter depth. The borehole is therefore not linked to direct rainfall recharge but is replenished from deeper lateral groundwater flow systems. Observation well W44 is located very close to a stream and is most likely influenced by stream flow.

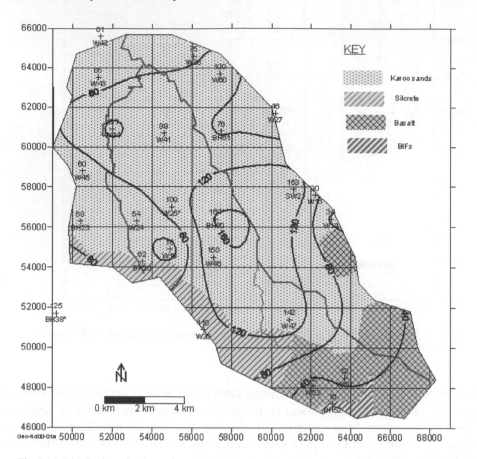

Fig. 7.14. Distribution of recharge from the WTF method in relation to geology in the Nyundo catchment.

Fig. 7.14 shows the spatial distribution of recharge (contours) in relation to the geology (shading) in the catchment.

Recharge is generally low in the basaltic regions to the southeast of the catchment and higher in the silcrete zone and Karoo sand areas.

The basaltic areas are overlain by silty clay red soils of low permeability and low specific yield. Recharge is therefore impeded. The sandy areas have higher permeability and specific yield resulting in more favorable recharge conditions. The silcrete zone has an even higher permeability than the sand and the recharge is correspondingly higher in this area.

The results show that on average the recharge in the sand areas is 1.5 to 2 times more than the recharge in the basaltic area. The recharge in the silcrete area is almost twice the corresponding recharge for areas at a similar altitude.

136

Conclusions

Tentative conclusions on the spatial variability of recharge from the WTF method are that the recharge in the catchment is influenced by topographical slope and geology but not by vegetation and land use. The recharge mainly occurs in the middle elevation areas of the catchment and not in the upper parts as is usually assumed. Recharge occurrence is by both preferential and diffuse flow and the recharge magnitude ranges between 2 and 28 % of annual rainfall. It is not only the amount of rainfall that determines the magnitude of recharge, the capacity of the aquifer to absorb and transmit the moisture plays a role. A threshold rainfall amount below which recharge is negligible exists.

7.4   Recharge with a spatial model: MODFLOW

For the Nyundo area a groundwater flow model has been developed and analysed with the MODFLOW to simulate the groundwater system and investigate the spatial distribution of recharge.

Sanford (2002), following the views of Winter (2001) argues that recharge is controlled by geology, climate or both and proposes that if model parameters are known well enough a groundwater flow model can be used to constrain both the rate of recharge and its distribution. Inverse modeling techniques are used to obtain, and hence calibrate, model output as a fit to some observable property of the groundwater system such as groundwater levels and water balance variables mainly flows (William, 1986). Once calibrated the model becomes a tool for solving common groundwater problems such as contaminant transport (Prommer et al. 2000), water resource depletion (Sophocleous & Perkins 2000) and irrigation drainage problems (Boonstra & Bhutta, 1996).

7.4.1 Model development

Table 7.13 summarizes the details of the model grid and conceptual parameters as developed for the application of MODFLOW.

Table 7.13. Model grid.

MODEL COMPONENT	MODELLING APPROACH	IMPORTANT CONSIDERATIONS	VALUES
Mesh Size.	Whole catchment with cells of uniform size.	Steep gradients likely to influence flow interchange.	100 m x 100 m
Layers	Layer A: soil zone, unconfined.		1 – 10 m thick
	Layer B: sub-divided into weathered and fractured rock zones.	Account for changes in layer thicknesses and spatial distribution.	5 – 30 m thick
	Flow exchange with Nyundo river.	Assumes permanent water flow is present at the river	-1
Boundary	Impermeable catchment boundary.	All cells outside catchment made inactive.	0
Layer Elevations	DEM from topographic surface (verified with GPS & tache field leveling)	sharp gradient changes near hills, river banks and gulleys.	Variable
	Bottom of layers: subtract average depths from topographic surface levels.	To capture spatial variations in lithology.	Variable

The model consists of two layers. First there is a layer a top layer A representing the soil zone, mainly the Karoo sands. The underlying layer B represents the weathered/fractured zone mix. Layer B has been sub-divided into two layers to account for the weathered and fractured zones in the geological model. Thus, the cell dimensions specified in Table 7.13 covers a quasi three layer model of 18 000 cells per layer based on the geological model of Section 5.3.

Table 7.14 summarizes the model parameters used in the flow model. A monthly time step has been selected to cut on processing time whilst meeting the requirement to assess available resources at a monthly time step. Also the calibration data from the manually monitored observation wells were at a monthly time step.

Table 7.14. Model parameters.

MODEL COMPONENT	MODELLING APPROACH	IMPORTANT CONSIDERATIONS	VALUES
Time	Transient state Stress period = time step	Water balance and groundwater level changes	1 mth
Initial hydraulic heads	End of dry season, September levels.		Variable
Boreholes and Observation wells	Observed heads used for calibration.	Match observed levels with modeled levels	Variable
Hydraulic conductivity	Field tests, literature & sensitivity analysis. Assume: $K_v = 0.5\,K_h$	Based on lithological variations	$K_h =$ 0.05 – 0.5 m/d
Effective porosity	Soil grading tests, literature, and sensitivity analysis.	Based on lithological variations	0.12 – 0.23
Specific yield	Literature and sensitivity analysis	Based on lithological variations	0.01 – 0.05

For the input parameters required to compute the flow exchange with the Nyundo aquifer and drainage to tributaries the limitations in data availability as well as the hydrology of the catchment had to be considered. River channel elevation has been derived from the river profiles of the DEM whilst river depths have been assumed to be 200 mm based on field observations. All ephemeral tributaries of the Nyundo river have been treated as drains with an average elevation of 1 m below ground surface.

To overcome the input of numerous wells with extremely low discharges representative wells have been used. Each representative well is located centrally to a demarcated demand area and the discharge estimated from the population served and the measurements from surrounding wells.

Initial estimates for net recharge have been used as input into the flow model thus eliminating the impact of interception, transpiration and capillary rise in the flow model. Recharge factors have been used to account for the impact of land use, terrain geometry and geology. The recharge for a given area has been estimated from Equation 7.31.

$$R_i = R_{wb} \cdot f_{Ri} \tag{7.31}$$

where, R (mm/month) denotes recharge, i is an area counter, R_{wb} is recharge from the water\balance model and f_R [-] is a recharge factor. Table 7.15 summarizes the recharge factors applied in the Nyundo catchment. The factors have been estimated on the basis of observed fluctuations in water levels. The recharge has been determined from the water balance model of Section 6.3. This approach offered a means to spatially distribute the recharge estimated from the lumped parameter estimate thereby offering a means of spatial comparison for the water balance estimate and estimates from the WTF and the CMB methods.

Table 7.15. Recharge factors.

ZONE	DESCRIPTION	R-factor	%Area	COMMENTS/REMARKS
1	pasture/flat land	1.2	20%	most natural/dominant system
2	pasture/slope	1.1	6%	slope favours runoff
3	cultivated/flat	1	30%	ponding with preferential paths
4	cultivated/slope	0.85	10%	runoff favoured
5	silcrete	1.1	11%	porous nature, high infiltration
6	acacia/miombo	0.95	1%	high interception
7	clay silt/slope	0.8	6%	low infiltration, runoff favoured
8	clay silt/flat	0.9	14%	runoff reduced
9	marshes	0.5	3%	discharge zone, but early

7.4.2 Results of MODFLOW analysis

The flow model has been calibrated on groundwater levels and river discharges. The calibration has been a staged process. In the first stage obvious model input and interpolation errors were removed or acknowledged. The second, and more crucial, stage of the calibration process has been the optimization of the recharge values

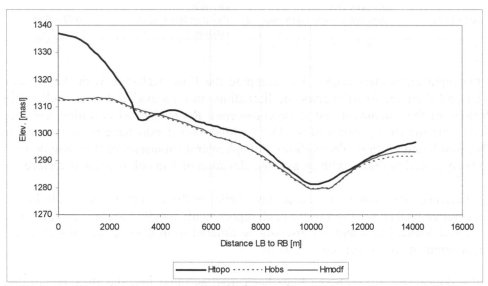

Fig. 7.15. Actual versus modeled groundwater levels at a cross-section of the Nyundo catchment.

Fig. 7.15 shows the variation in observed and calculated water levels for a cross section in the middle of the catchment for stress period 26 (February 2001). The flow model approximates the groundwater levels correctly across the catchment except near drainage channels. Terrain affects the estimated water levels for both interpolations from groundwater observations and for model calculations. Sharp changes in slope at river channels and gullies result in water tables well above the actual ground levels.

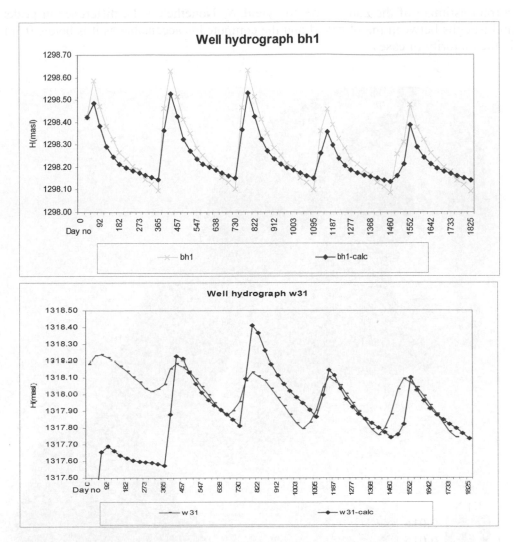

Fig. 7.16. Modeled versus observed temporal water table fluctuations in the Nyundo catchment.

Fig. 7.16 shows the temporal variations in the water table for two wells from the catchment over the four-year simulation period. Well bh1 is close to a natural drainage stream whilst w31 is higher up in the interfluves. (See also Fig. 5.2 and note that the mismatch at the beginning of well 31 is due to the difference in initial conditions between the model and the field observations). The figure shows that it is largely possible to correctly simulate the trend in the water table fluctuations for these wells. The recession curve is correctly simulated for the well near the natural drainage (bh1) suggesting that there is hydraulic connectivity between the groundwater system and the surface flow.

The model computations for the higher well (w31) suggests a delayed and less direct link with the surface drainage. The delay could be due to heterogeneity in aquifer parameters as well as uncertainties in lithological boundaries or an incorrect recharge distribution over time. Thus, though feasible, it is not always possible to develop a correct geological model of the crystalline basement aquifer. The used model is an approximation rather than a correct replica of the ground reality.

Peaks and troughs of the modeled and observed groundwater hydrograph coincide in time but not in magnitude highlighting the difficulties associated with determining a correct estimate of the aquifer specific yield, S_y. Nonetheless the difference in peaks and troughs between the observed and the computed is acceptable as it is below 0.5m in the majority of cases.

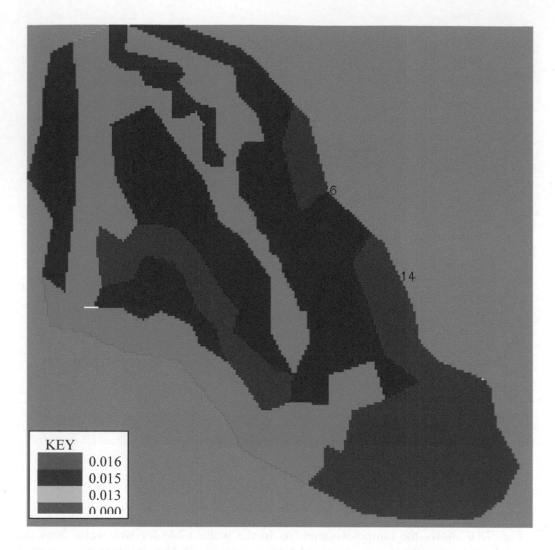

Fig. 7.17. Recharge map (m/month) used in MODFLOW for November in the year 2000/01.

Fig. 7.17 shows a recharge map derived from the MODFLOW model for the hydrological year 2000/01. the recharge is largest in cultivated areas and those areas dominated by loose grained lithologies. (See Section 5.3). However, as the recharge is imposed through an inverse[26] modeling procedure the derived recharge does not reflect the actual dynamics in the area. Second, as recharge zones are predetermined, overlapping zones are not fully accounted for.

---

[26] The recharge is assumed and taken as model input data to be checked against the flow parameters such as groundwater levels and the water balance.

7.4.3 Conclusions on MODFLOW

Though an important tool for groundwater flow analysis, MODFLOW has limitations when applied in areas of with uncertainties in aquifer geometry such as the Nyundo aquifer. Uncertainties in the aquifer geometry imply that a correct hydrogeological model will only be partially achieved. To do so will require considerable investments in field data collection (geophysical investigations, in particular accurate lithological information from drilled wells in the study, and water balance measurements) and processing time. Both require financial resources, a commodity rather scarce for most developing nations.

The misgivings above aside however, MODFLOW can be of some use in investigating the distribution of recharge and when used in combination with a lumped parameter water balance model may provide a means for determining regional water resources. In the case of the Nyundo MODFLOW showed that the groundwater flow in the catchment is localized rather than regional and that a two geological layer model is sufficient to describe the groundwater system in crystalline basement aquifers. The use of the drain package to simulate ephemeral streams strengthened the argument that stream discharge is generated by through-flow rather than direct surface runoff.

7.5   A word on the recharge estimation methods

The different recharge estimation methods have different strengths and weaknesses and perform differently at different time scales. The WB method explains better the occurrence of recharge and can give estimates at a monthly time step. However it cannot give the spatial distribution of recharge. The CMB method only applies at a long time scale, minimum of a year and is very sensitive to the chloride deposition estimate. The WTF on the other hand can explain the dynamics of recharge even at a daily time step. MODFLOW is useful in confirming the distribution of recharge when used in combination with a water balance model but relies on the correct assumptions on the hydrogeological model.

CHAPTER EIGHT

8.   SUMMARY AND SYNTHESIS

This chapter presents the main findings, conclusions and propositions of the study. The findings on catchment characterization, water balance and recharge estimation are presented. The performance of the different recharge estimates in space and time are analyzed and the development of monthly models for recharge estimation argued.

8.1   Conceptual models for CBAs

The hydrological model

Groundwater recharge estimation models used in the region seldom consider the rainfall storm as the starting point for determining groundwater recharge since most of the methods are based on time steps longer than the time scale of an ordinary storm. In fact most methods seek to determine the annual recharge. Similarly with respect to space, the recharge estimate is often required for a region whose spatial extent is greater than that of a single storm.

Analysis of the rainfall data presented in Section 5.2 of this discourse shows that rainfall storms are predominantly of short duration, high intensity and limited spatial extent. While the time scale of recharge (days to a month in the Nyundo catchment) is larger than the one for rainfall storm the pattern of rainfall (frequency, storm extent) determines the amount of recharge as it controls the rainfall amount that does not become recharge.

Storms are convective, i.e., short duration, high intensity and limited spatial extent. The combination of high temperature and short storm duration enhance the importance of interception. As such interception is the single most important rainfall component in determining the likelihood of recharge.

The rainfall in the region exhibits Markov properties (See Section 5.2). The interception models by De Groen (2002) can be adopted for determining effective rainfall.

The areal extent of tropical storms is less than 10 km which is less than the size of most catchments for which recharge is required. Thus in a given catchment a series of spatially distributed storms will determine the average catchment recharge. Standard methods of determining areal rainfall can be applied if the rainfall monitoring network is high enough. Otherwise for rapid assessment purposes representative point values of rainfall can be used.

The water balance modeling, rainfall data analysis and groundwater flow modeling exercises of this discourse suggest that rainfall seasonality is important. The recharge has been shown to follow the rainfall part and the recharge efficiency (R/P) changes as the season progresses. The recharge/discharge relationship in a catchment follows the rainfall seasonality. Recharge occurs in the wet season and discharge is predominant in the dry season as baseflow in natural drainage systems. For resource assessment purposes it can be postulated that at a long time scale;

$$\text{Recharge} + \text{discharge} = 0 \qquad\qquad (8.1)$$

Equation 8.1 implies no accumulation of groundwater from one hydrological year to another. Such a management approach considers only the renewable groundwater resource and also unconsciously guarantees a form of groundwater environmental reserve since the water contained in the parts of the aquifer not hydrologicaly connected to the natural drainage system is not exploited.

The hydrogeological model

The spatial heterogeneity of crystalline basements and imprecise nature of geophysical measurements make it difficult to be certain of a chosen configuration for a hydrogeological model. The three layer conceptual scheme, consisting of a soil layer overlying a decomposed/weathered and a fractured layer below, proved difficult to model. Whilst the model is fairly dominant in the geophysical measurements at several points, its areal representativeness remains suspect in the groundwater flow model. The reason for this could be that the actual heterogeneity in the crystalline basements in such a model is based on point geophysical measurements cannot adequately describe the field reality. As such any model based on point measurements only results in a "dummy" of the field reality. The heterogeneity of the dummy does not equate to that of the field reality. Hence, the modeled groundwater flow is at variance with field groundwater flow to a degree. The more layers are assumed for the hydrogeological model the greater that variance since more data points (and/or parameters) are required to adequately describe the groundwater system.

The three layer hydrogeological model discussed in Section 1.1 is correct in defining the structure of the CBA but is not useful in assessing the available renewable groundwater resources. The usable part of the CBA is the soil/weathered zone that is recharged directly by rainfall and is hydraulically connected to the natural drainage system. The fractured zone in the three layer model can be considered as a groundwater environmental reserve which will be exploited only in times of water stress such as droughts.

Consequently for groundwater assessments, the three layer model can be reduced to a simpler two layer system with fewer parameters that is easier to manipulate and results in more "realistic" groundwater response pattern (Sekhar et al., 1994).

The water balance model for a CBA

The lumped parameter approach applied to a simple reservoir system can reasonably simulate the water balance of a crystalline basement aquifer system.

The following are some of the main conclusions from the water balance model of a crystalline basement aquifer in a semi-arid tropical area.

Up to 90% of annual rainfall falls in a single wet season and potential evaporation is higher than total rainfall. Consequently groundwater recharge estimation methods based on the water balance are only applicable if taken at short time steps during which time the rainfall can be higher than evaporation.

Direct surface runoff is negligible and much of the flood flow is generated from subsurface flow. The spatial distribution of tropical storms is limited such that modeling stream discharge at a daily time step requires a relatively dense rainfall monitoring network so that a reasonable flow gauge to rain gauge ratio is achieved. A one to one ratio does not correctly simulate the hydrological behavior due to the mismatch between rainfall records and flow observations.

Actual transpiration is affected by antecedent soil moisture conditions even at an annual time step. For example, modeling results show that a wet antecedent year results in low transpiration in the current year. Interception range of 4-6 mm/d best describes the water balance in the catchment but the method of calculation of the interception has a bearing on the estimates.

Groundwater recharge is by preferential and diffuse flow mechanisms. The root depth governs the transpiration from the catchment. The groundwater flow is best simulated by considering two types of flow – a fast flow component with a time scale of 3-5 days and a slow flow component with a time scale of 90-120 days. Annual estimates of recharge depend on the estimates of interception, evaporation and transpiration.

Water balance modeling has shown that though linear mathematical expressions are feasible at monthly and/or annual time steps such expressions depend on the occurrence and nature of the process for which the expression is required. The modeled process needs to be divided into its components and the time scale of each component considered separately to derive component specific expressions. For example, in this study it was not possible to simulate correctly the catchment stream discharge until the stream discharge was split into a fast component and a slow component. Water balance modeling attempts of this study also show that the spatial heterogeneity of rainfall makes it very difficult to model at a daily time step.

8.2   Comparisons of groundwater recharge estimation techniques

In this section the recharge methods used in the study are compared. How accurate is each method in detecting the occurrence of recharge? How accurate the estimated magnitude at monthly and annual time steps? How correct are the predictive estimates of recharge in time (and space)?

Occurrence of recharge

When does recharge occur?

The WTF and WB methods at daily time steps give clues to this question more than the CMB method which only applies at longer time steps, usually a year.

Groundwater recharge from rainfall occurs mainly between December and February though it can be observed as early as November with the earliest rains if the rainfall depth exceeds 5 mm/d. There is an annual threshold value of rainfall, around 250 mm/a, below which recharge is minimal. This threshold can be attributed to interception. Preferential flow is evident and can be as high as 80% of the recharge due to a rainfall event. There are daily thresholds for preferential recharge. Daily rainfall amounts below 5 mm/d do not cause recharge.

The occurrence of recharge also depends on the magnitude of the preceding annual, monthly and daily rainfall as well as the magnitude of the recharging rainfall. These determine whether recharge is predominantly preferential or diffuse flow. Dry antecedent soil moisture conditions tend to favor preferential flow whilst wet antecedent conditions enhance diffuse flow.

146

In a dry year soil moisture deficits limit diffuse flow since field capacity conditions are not attained. Preferential flow is more likely to occur because of more voids and cracks due to soil shrinkage, drying (or dying) vegetation roots, widened animal/insect burrows etc. This observation is important in understanding the availability of groundwater resources in the shallow crystalline basement aquifers since it suggests that the only source of recharge in years of low rainfall may be preferential flow. If correct, this is of vital importance for water resources management in CBAs.

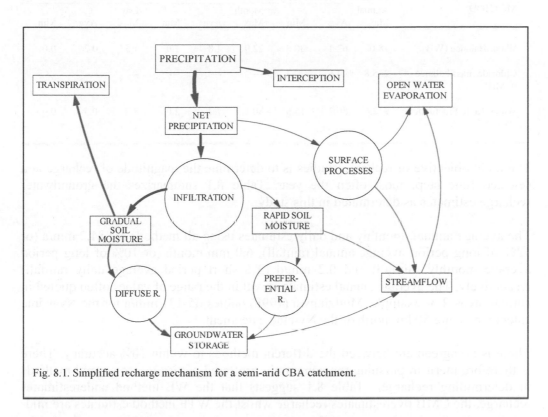

Fig. 8.1. Simplified recharge mechanism for a semi-arid CBA catchment.

In light of the above, Fig. 2.1. in Chapter Two which shows the various elements of recharge in a semi-arid area, can be simplified as shown in Fig. 8.1 to describe the recharge/discharge processes in a CBA such as the Nyundo catchment. The figure suggests that to better understand recharge we need to understand other processes in greater detail, mainly the interception, transpiration and infiltration. Hence, there is merit in studying landuse as a starting point in recharge studies.

There are two mechanisms of recharge in CBAs not three, namely diffuse and preferential flow recharge. Indirect recharge from stream flow is negligible. Infiltration is a dominant characteristic at the catchment scale as it controls not only soil moisture potential but the generation of stream discharge as well. A theory is advanced here that soil moisture develops in two forms. Rapid soil moisture build-up results in saturation conditions in the top part of the soil profile. The excess moisture is released through preferential paths to the deep groundwater storage or results in fast groundwater seepage to the stream discharge channels. Gradual soil moisture build up occurs through the percolation of infiltrated water to the saturated zone. Part of the percolating water is intercepted by transpiration thus reducing the net downward flow.

Surface processes are not a dominant factor in the recharge process or the generation of stream discharge. Localised recharge from surface ponding is therefore negligible. Only extreme rainfall will generate surface runoff. Note however that this

147

phenomenon is most likely so because most soils in the area are sandy and have a high infiltration capacity. The generation of surface runoff will increase with an increase in the clay content of the soil.

The magnitude of recharge

Table. 8.1. Catchment average groundwater recharge estimates from different methods at different time steps.

METHOD	ESTIMATE (mm/time unit)								
	Annual			Month			Day		
	Max	Avg	Min	Max	Avg	Min	Max	Avg	Min
Water Balance (WB)	78.0	62.1	30.8	22.9	4.8	1.0	1.5	0.2	0.0
Chloride mass balance (CMB)	235.8	94.6	39.3	---	---	---	---	---	---
Water Table Fluctuation (WTF)	212.5	86.2	15.6	30.1	7.2	2.0	8.2	0.5	0.0

The main objective of recharge studies is to determine the magnitude of recharge at a selected time step, most often the year. Table 8.1 summarizes the groundwater recharge estimates as determined in this study.

The average annual, monthly and daily estimates using all methods are 84.2 mm/a (or 12% of long period average annual rainfall), 6.0 mm/month (or 10% of long period average monthly rainfall) and 0.2 mm/d (3% short period average daily rainfall) respectively. The average annual estimate is within the range of values often quoted in similar areas. For example, Mudzingwa (1998) quotes 65-131 mm/a for the Nyatsime catchment some 50 km north of the Nyundo catchment.

There is no agreement between the different methods to within 10% accuracy. There is therefore merit in pursuing the commonly proffered advice to use several methods in determining recharge. Table 8.1 suggests that the WB method underestimates recharge, the CMB overestimates recharge whilst the WTF method estimates are mid-range.

Fig. 8.2 shows the correlation between the CMB and WTF methods. The scatter plot shows than the estimates are in agreement for values below 150 mm/a or 20% of the long period average rainfall. Beyond this 20% level recharge is excessive and very localized. WTF may capture such values but the CMB method will not. Plots of the recharge estimates against land elevation and depth to water table shows that the two methods are generally in agreement and are not affected by elevation or depth to the water table.

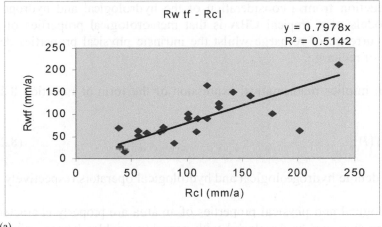

(a)

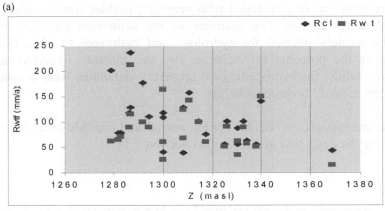

(b)

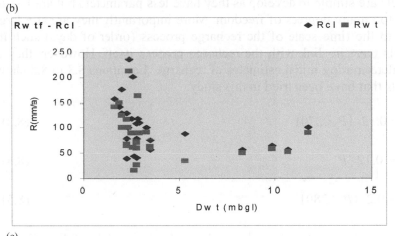

(c)

Fig. 8.2. Correlation between R_{cl} and R_{wtf} (a) scatter plot, (b) wrt elvation and (c) wrt depth to water table.

8.3   Groundwater recharge estimation models for CBAs

The main deduction from a consideration of the hydrological and hydrogeological conceptual models of a tropical CBA is that meteorological properties of an area govern the occurrence of recharge whilst the intrinsic physical properties determine the magnitude of recharge.

This deduction implies mathematical expression of the form of Equation 8.2 can be used.

$$R \approx F[f(P)] \qquad (8.2)$$

where F and f denote hydrogeological and hydrological operators respectively.

Thus if hydrological and physical properties of an area are properly characterized an empirical expression can be developed with generic variables whose values can be driven from those properties. The accuracy of the estimation equation therefore depends on the models selected for hydrology and hydrogeology. The hydrology model determines the potential for recharge, i.e., the amount of rainfall that may become recharge whilst the hydrogeological properties determine what proportion of the potential may actually become recharge.

Such mathematical models are feasible for the annual and monthly time steps but are not applicable for the daily time step (See also Section 7.1).

Annual models

Annual models are simple to develop as they have less parameters but are less reliable as they have too many degrees of freedom. More importantly they are at a large time step relative to the time scale of the recharge process (order of days) such that the annual models have no link with the recharge process itself. However, they remain valuable for determining initial estimates of recharge. Equations 8.3 to 8.5 show three annual models that have been tried in this study.

$$R_{wb} = 0.12 \cdot (P - 270) \qquad (8.3)$$

$$R_{cmb} = 0.12 \cdot P \qquad (8.4)$$

$$R_{wtf} = 0.2 \cdot (P - 280) \qquad (8.5)$$

Table 8.2 shows the results of applying the recharge models of Equation 8.3 and Equation 8.4. The table shows that annual recharge models give a range of between 6% and 14% of annual rainfall. The water balance derived recharge model under estimates recharge but the water table fluctuation model over estimates recharge when compared with the recharge average derived directly from the field estimates.

Table 8.2. Comparison of annual recharge models (mm/a).

Year	P	R_{wb}	R_{cmb}	R_{wtf}
1998/99	743.5	56.8	89.2	92.7
1999/00	961.9	83.0	115.4	136.4
2000/01	945.3	81.0	113.4	133.1
2001/02	508.3	28.6	61.0	45.7
4 yr average		62.4	94.7	102
Deviation		-21.8	10.6	17.8

*Deviation from the average recharge R_{avg}.

Monthly models

Monthly models give a better simulation of the recharge process as they are closer to the time scale of the recharge process. Equations 8.6 is the general case of a multiple linear regression model used to estimate the monthly recharge in this study.

$$R_0 \cong b_0 \cdot MAX(P_0 - I, 0) + b_1 \cdot MAX(P_1 - I, 0) \dots + b_n \cdot MAX(P_n - I, 0) \qquad (8.6)$$

Where R, P and I denote recharge, rainfall and interception respectively and b is a recharge coefficient. The subscripts 1 to n denote the months preceding the month for which recharge is being estimated (denoted by 0 in Equation 8.6).The equation assumes that the recharge in a given month is partly from the rainfall of that month and the soil moisture due to the rainfall of the preceding months. Interception threshold is assumed to be constant for all months. It is essential to note that actual interception is not constant, only the threshold is. This also explains the need to use the MAX operator in Equation 8.6.

When applied to Nyundo data Equation 8.6 reduces to Equation 8.7.

$$R_0 \cong 0.13 \cdot (P_0 - 50) + 0.03 \cdot (P_1 - 50) \qquad (8.7)$$

Equation 8.7 suggests that the monthly interception is around 50 mm/month and that less than 20% of recharge is due to moisture stored from the previous month. Only two months influence recharge.

The handicap of a constant monthly interception can be removed if monthly interception is determined directly from rainfall. De Groen (2002) developed monthly models for interception and transpiration as functions of rainfall depth and rainfall occurrence. These models have been tested against the results of the daily water balance of this thesis.

Fig. 8.3 summarizes the correlations between the daily water balance interception and the De Groen (2002) monthly interception model estimates. The monthly interception model gives good estimates of the monthly interception with a coefficient of determination of 0.97 and a proportionality constant of 1 between the water balance value and the monthly interception model estimate. The monthly model therefore, can be applied to the Nyundo catchment to estimate the monthly interception and determine the net rainfall.

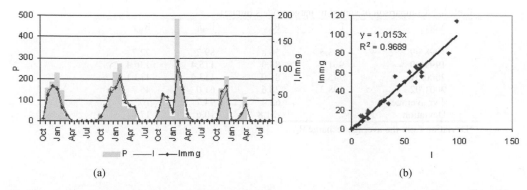

Fig. 8.3. Interception from the daily water balance and monthly interception model: (a) time series fit and (b) linear relation.

On the basis of the above observations a monthly recharge estimation model has been developed. Since the correlation between the water balance interception and monthly interception model is strong, the net monthly rainfall estimate of the interception model has been adopted. Equation 8.8, after De Groen (2002) gives the net monthly rainfall estimated from the monthly interception model.

$$P_{net,m} = P_m \cdot \exp\left(\frac{-I_p \cdot n_r}{P_m}\right) \qquad (8.8)$$

where P_m (mm/month) is the gross monthly rainfall, I_p (mm/d) is the daily interception threshold and n_r [-], is the number of rain days in the month.

In the original De Groen equation for transpiration a coefficient, B, defined as the rate at which transpiration changes with net rainfall, is used as a constant of proportionality between the transpiration and the net rainfall. Where surface runoff is negligible the net rainfall is split between transpiration and recharge. The coefficient B defines the proportion of net rainfall that contributes to transpiration. In such a case the proportion of net rainfall contributing to recharge is the difference between unity and the transpiration proportion. The monthly recharge model is therefore given by Equation 8.9.

$$R_m = (1 - B) \cdot P_m \cdot \exp\left(\frac{-I_p \cdot n_r}{P_m}\right) \qquad (8.9)$$

The daily water balance model results show that transpiration accounts for 88% of net rainfall. A value of 0.88 has therefore been adopted for the constant B in Equation 8.9.

Fig. 8.4 below shows the performance of the models of Equations 8.7 (graph Rlin2) and Equation 8.9 (graph R2mmg) over a four year period against the average monthly recharge determined from water balance calculations (R) and the observed stream discharge (Qobs).

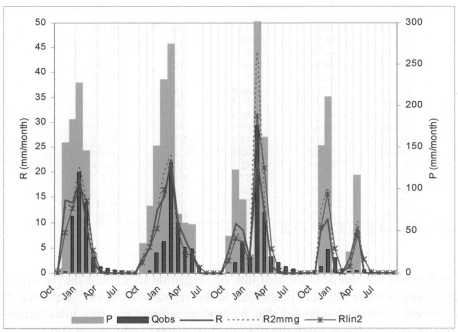

Fig. 8.4. Monthly groundwater recharge estimates for a rain season from different estimation models.

The recharge models and water balance estimates generally agree. In the first half of the rainfall season discharge occurs about a month after rainfall indicating moisture build-up. In the second half recharge and discharge occur in the same month.

Equation 8.9 estimates the fraction of net monthly rainfall that is likely to be recharge. Following the concept of Equation 8.2, the coefficient ($1-B$) is a property of the aquifer material which determines the amount of net rainfall that contributes to recharge. From soil mechanics theory it can be argued that the water that can be absorbed and drained from the soil is that water held in the pore spaces and not abstracted by roots through plant transpiration. The coefficient ($1-B$) can therefore be assumed to be determined by the effective porosity, ϕ [-], root depth, D_r [L] and depth to the water table, D_{wt} [L] as in Equation 8.10.

$$(1-B) = \phi \cdot \left(1 - \frac{D_r}{D_{wt}}\right) \tag{8.10}$$

The coefficient ($1-B$) is the sum of the coefficients b_0 to b_n in Equation 8.6 and will be equated to $0.13 + 0.03 = 0.16$ in Equation 8.10.

Table 8.3 shows the variation in the coefficients ($1-B$) and b_1 with respect to the depth to water table for wells in karoo sands in the Nyundo catchment.

Table 8.3. Model recharge coefficients from field data.

ID	X	Y	H	D_{wt}	(1-B)
**B2	57700	56500	1294.3	2.91	0.114
BH30	57220	56780	1300	2.87	0.112
W44	52030	60910	1282.9	2.84	0.110
W45	50530	58780	1300	2.82	0.109
W53	62110	48130	1300	2.82	0.109
BH52	63080	47230	1368.8	2.67	0.098
W42	51380	65580	1279.5	2.61	0.093
W41	54620	60700	1286.5	2.58	0.091
W28	56080	64560	1286.9	2.42	0.077
W25	54970	57050	1314.5	2.35	0.070
BH23	50450	56280	1308.1	2.33	0.068
W43	51290	63490	1282.4	2.33	0.068
BH61	57490	60800	1287.1	2.32	0.067
BH38	49200	51730	1308	2.09	0.042
W60	57370	63650	1291.8	2.05	0.037
W46	57100	54470	1340	1.91	0.017

Lithology = karoo sand, ϕ = 0.33, D_r = 1.8 m,

Equations 8.6 and 8.9 can be combined into one equation to account for the effect of interception and aquifer physical properties in the recharge estimate. Equation 8.12 gives a general combination model that incorporates direct monthly interception as well as the delay in recharge using the physical parameters as discussed above.

$$R_0 = \phi \cdot \left(1 - \frac{D_r}{D_{wt}}\right) \cdot P_0 \cdot \exp\left(\frac{-I_p \cdot n_{r0}}{P_0}\right) \qquad (8.11)$$

Considering average values from Table 8.3 the sum of the coefficients b_n can be equated to 0.10 whilst interception can be taken as 5 mm/d from the water balance modeling results. Thus, the recharge estimation model for the Nyundo catchment can be given by:

$$R_0 = 0.10 \cdot P_0 \cdot \exp\left(\frac{-5 \cdot n_{r0}}{P_0}\right) \qquad (8.12)$$

Table 8.4 summarizes the recharge estimates for the different models against the recharge determined from the water balance and water table fluctuations for the hydrological years 1998/9 to 2000/1.

Table 8.4 shows that the model of Equation 8.13 (R_{mod} in Table 8.4) gives recharge estimates accurate to within ± 10% of both the water balance (R_{wb}) and the WTF method (R_{wtf}).

Table 8.4. Monthly recharge estimates from models (mm/month).

MONTH	1998/1999			1999/2000			2000/01			2001/2002		
	R_{wb}	R_{wtf}	R_{mod}	R_{wb}	R_{wtf}	R_{mod}	R_{wb}	R_{wtf}	R_{mod}	R_{wb}	R_{wtf}	R_{mod}
October	0.0	0.0	0.1	1.6	2.1	2.9	1.8	2.4	2.6	0.0	0.0	0.0
November	6.2	8.0	11.3	4.9	5.1	5.8	6.4	7.0	8.0	7.0	9.0	10.6
December	13.5	12.8	13.6	8.7	8.8	10.8	7.3	6.0	5.3	16.0	15.6	16.1
January	17.5	17.1	20.1	15.9	16.4	19.0	3.2	2.2	1.3	9.0	4.8	1.8
February	16.8	14.2	14.1	21.8	20.5	22.8	24.5	31.2	39.2	0.0	0.2	0.0
March	7.3	4.5	2.8	13.0	8.3	6.2	29.8	20.8	15.7	0.9	1.2	1.2
April	0.6	0.4	0.0	3.8	3.8	3.2	7.3	4.9	0.6	6.3	7.4	9.1
TOTAL	61.9	57.0	62.0	69.7	65.0	70.7	80.3	74.5	72.7	39.2	38.2	38.8
	8.3	7.7	8.3	7.2	6.8	7.4	8.5	7.9	7.7	7.7	7.5	7.6

However, in most cases daily data are not available and it may be necessary to determine an average monthly interception. It is suggested here that transitional probabilities for rainfall occurrence be used.

De Groen (2002) showed that the number of rain days in a month can be estimated from probabilities of daily rainfall occurrence. A simplified estimation formula is given by:

$$n_r = \frac{30 p_{01}}{0.66}$$

(8.13)

Where p_{01} is the probability of a wet day after a dry day. Further elaboration of the Equation 8.13 yields Equation 8.14.

$$n_r \cong P^{0.55}$$

(8.14)

Where P (mm/month) is the monthly rainfall.

Equation 8.12 can now be re-written as:

$$R_0 = \phi \cdot \left(1 - \frac{D_r}{D_{wt}}\right) \cdot P_0 \cdot \exp\left(-I_p \cdot P_0^{-0.45}\right)$$

(8.15)

Application and limitations of model

The monthly model has been develop and tested on Nyundo data. The Nyundo terrain is undulating and slope effects can be neglected. The time scale for the recharge is two months and no distinction is made between preferential flow recharge and diffuse flow recharge. It can be expected that in places where the weathered zone is more pronounced as compared to the soil zone preferential flow recharge will be more dominant thus reducing the time scale for the recharge process.

The depth to the water table in the Karoo sand of the Nyundo is relatively constant and depth differences do not exceed 50%. Areas of distinct lithology and land use (concept of hydrotopes) can be identified in the Nyundo catchment as such it is easy to identify soil and root depths. Difficulties are to be anticipated in highly heterogeneous areas.

To test the applicability of the model in areas outside the Nyundo catchment rainfall and physical data for selected crystalline basement areas have been used to estimate recharge and compare with results from other researchers. Long period (20 years) average monthly rainfall data has been used. Physical data is derived from general soils, hydrogeology and vegetation maps of the selected areas. Potential interception is set at 5 mm/d in line with the arguments of this discourse.

Table 8.5a: Comparison site characteristics.

Location	ϕ (-)	D_r (m)	D_{wt} (m)	I_p (mm/d)	Description
Nyatsime	0.33	1.0	2.5	5.0	Grass and tree mix, sandy soils
Marondera Grasslands	0.33	0.5	1.5	4.5	Grass only, sandy soils
Chiweshe	0.33	0.7	2.0	5.0	Grass and tree mix with rock outcrops, sandy soils

Table 8.5b: Comparative recharge estimates.

Location	Oct	Nov	Dec	Jan	Feb	Mar	Apr	Year	Other	Method	Reference
Nyatsime	2.0	9.2	16.1	25.0	18.3	7.3	1.5	77.9	162.0	Flux analysis	Mudzingwa &
									74.0	Reservoir	Lubezynski
									130.0	WTF	(1999)
Marondera Grasslands	4.4	12.2	24.8	30.8	22.3	12.2	5.1	106.6	185.0	CMB	Jarawaza
									190.0	WTF	(1999)
									136.0	CMB	McCartney (1998)
Chiweshe	2.2	8.6	22.2	23.0	18.7	8.6	2.7	83.2	71.6	CMB	Mjanja (2000)

Table 8.5 shows the performance of the model when applied to areas similar to but outside the Nyundo catchment. The model yields recharge estimates comparable to those from other methods.

A word of caution

The model of this discourse does not seek to replace direct recharge estimation methods but aims to assist the water resources manager to make an initial evaluation of available groundwater resources upon which decisions on water allocation and management may be based.

Cost is a major consideration and constraint for effective water resources management for poor regions of the world. Often, rules of thump are developed to avoid the costs associated with full hydrological investigations. For example, in Zimbabwe recharge was at some time estimated as "2 % of annual rainfall" when determining the groundwater to be allocated to a permit. The rapid assessment model of this discourse may go along way in avoiding such rudimentary estimates.

With reasonable data on soil properties and depths to water table simple groundwater recharge monographs can be derived from the recharge model and may be easily applied in water resources assessments without the need for technical expertise and massive capital investments.

CHAPTER NINE

9.   CONCLUSIONS

9.1   General conclusions

Water balance

The annual rainfall in the Nyundo catchment and surrounding areas is seasonal, averages 730 mm/a and can be cyclonic or convectional. The rainfall storm occurrence is sporadic and of limited spatial extent.

Annual potential evaporation is higher than the annual rainfall thus potential evaporation cannot be used to estimate groundwater recharge at an annual scale. However, at a day time scale evaporation is constrained to a maximum threshold of 6 mm/d whilst rainfall can be higher than this threshold. Thus a daily water balance can be used to estimate groundwater recharge.

Actual evaporation can account for 80-95% of annual rainfall in a tropical environment. Interception alone can account for 25-35% of annual rainfall.

A threshold annual rainfall of 250-300 mm/a is required to meet evaporative demand before significant groundwater recharge (and hence stream discharge) is observed.

Stream discharge is generated though soil moisture excess rather than surface processes.

The aquifer system

Though VES resistivity profiles and borehole log sections show that a three layer geological model fits the aquifer configuration, a two layer model with a soil zone overlying a weathered/decomposed mantle is sufficient for flow modeling purposes.

It needs to be emphasized however, that resistivity profiling is not an adequate tool for determining the distribution of lithologies though it can give a reasonable impression of the aquifer geometry. This is so because readings are usually taken at a spatial scale bigger than that of the slope changes within the buried layers.

The water table for the Nyundo catchment is shallow with an average depth to water table of 2-5 m. The response of the aquifer to rainfall events, expressed through water levels, becomes more pronounced as the water table becomes shallower.

Recharge estimation

Three methods have been used to estimate both the amount and distribution of groundwater recharge in the Nyundo catchment, the chloride mass balance (CMB), the daily catchment water balance (WB) and the water table fluctuation (WTF) methods. The methods give similar but not the same amounts and distributions of groundwater recharge. All methods give a high spatial variability in recharge with coefficients of variation of 60-65%. Thus single point measurements of recharge are not good indicators of regional recharge. Recharge estimate for the Nyundo catchment is between 8% and 15% of annual rainfall.

The CMB method gives an upper limit of recharge and can only be used to estimate annual values. The chloride deposition, c_p, which is controlled by the origin of rainfall (recycled or straight from the ocean) determines the recharge estimate. Geology and land use have secondary influence. The method is not applicable in discharge areas.

The WTF method can show the magnitude of event recharge and its relationship with rainfall, the residence time of groundwater in the aquifer and the total annual recharge but is heavily reliant on the selected value of the specific yield. In the Nyundo catchment the method gives groundwater residence times of 100-120 days (3-4 months).

A daily lumped parameter water balance gives a lower limit of recharge and correctly identifies when recharge occurs but is sensitive to the assumptions on soil available moisture, interception and transpiration rates. The method is also negatively affected by the problem of high spatial heterogeneity of daily rainfall. The method agrees with the water table fluctuation on the occurrence of recharge. Recharge occurs mainly in February and is approximately 10-15% of annual rainfall.

The groundwater recharge is higher in the Karoo sand dominated and less rugged lower parts of the catchment rather than the more rugged weathered/decomposed mantle dominated high ground. The catchment exhibits the characteristics of a mixed plutonic and alluvial province.

MODFLOW provides difficulties when applied to crystalline basement terrain such as the Nyundo catchment due to the uncertainty in determining the layer boundaries and their topography. The rugged nature of the weathered/decomposed mantle cannot be ascertained using resistivity and borehole log data alone. Second, the degree of weathering has high spatial variability making it impossible to assign uniform properties to the same layer. Interpolation may not be the best solution given that the degree of weathering changes randomly in space.

Zoning the aquifer and assigning uniform properties per zone proved useful but even after this simplification what is modeled in MODFLOW is not the original aquifer but a (proxy) similar aquifer that mimics the real aquifer.

MODFLOW is useful in analyzing the saturated zone balance and thus validating the conceptual model especially for the Karoo sand dominated lower parts of the catchment. Of importance is its ability to estimate/validate direct outflows from the saturated zone particularly the effect of vegetation through capillary rise.

9.2   A monthly model for groundwater recharge estimation.

A combination of multiple regression and interception models suggest that it is feasible to express the monthly recharge estimate as a function of its influencing factors; rainfall (number of rain days, average monthly depth), daily interception threshold, transpiration impacts (root depth), lithological properties (effective porosity), terrain aspects (topographic slope) and state of the groundwater system (depth to water table).

The basic groundwater recharge estimate is obtained from:

$$R_0 = \phi \cdot \left(1 - \frac{D_r}{D_{wt}}\right) \cdot P_0 \cdot \exp\left(\frac{-I_p \cdot n_{r0}}{P_0}\right) \tag{8.11}$$

In the absence of daily data such as number of rain days, the monthly recharge model can be simplified to:

$$R_0 = \phi \cdot \left(1 - \frac{D_r}{D_{wt}}\right) \cdot P_0 \cdot \exp\left(-I_p \cdot P_0^{-0.45}\right) \tag{8.15}$$

The monthly model is only valid under the following assumptions:

The time scale for the recharge is around a month and no distinction is made between preferential flow recharge and diffuse flow recharge,

The terrain is undulating and slope effects can be neglected,

The depth to the water table is relatively constant and depth differences do not exceed 50%, and,

Areas of distinct lithology and land use (concept of hydrotopes) can be identified.

The monthly model only provides a rapid assessment estimate for management decision-making and cannot be taken as a substitute for direct methods for estimating recharge.

Model estimates must be quoted with a ±20 % error margin to give a recharge range rather than an absolute figure.

REFERENCES

Abdulrazzak, M. J., Sorman, A. V. & Abu-Rizaiza, O., 1988. *Estimation of natural groundwater recharge under Saudi Arabian arid climatic conditions.* In: Simmers, I. (Ed), 125-139.

Adar, E. M., Neuman, S. P. & Woolhiser, D. A., 1988. Estimation of spatial recharge distribution using environmental isotopes and hydrochemical data. I. Mathematical model and application to synthetic data. J. Hydrol. 97, 251-277.

Adar, E.M. and C. Leibundgut (Eds), 1995. *Application of tracers in arid zone hydrology.* IAHS Publication No. 232: 452 pp.

Allison, G. B. & Hughes M. W., 1983. The use of natural tracers as indicators of soil water movement in a temperate semi-arid region. J. of Hydrol. 60, 157-173.

Anderson MP & Wossner WW, 1991. *Applied groundwater modeling: simulation of flow and advective transport.* Academic Press, San Diego, CA. 381 pp.

Appelo, C.A.J. & Postma, D., 1999. *Geochemistry, groundwater and pollution.* Balkema.

Askew, A., 1996. *Network Design.* IHE Delft Lecture Note, HH267/96/1. 80pp.

Athavale, R. N., 1985. Nuclear tracer techniques for measurement of natural recharge in hard rock terrains. In: Proc. International Workshop on rural hydrogeology & hydraulics in fissured basement zones. Dept. of Earth Sicences, Univ. of Roorkee, Roorkee, India.

Barnes, J.W., 1995. *Basic Geological Mapping.* (3rd). John Wiley & Sons.

Bartram, J. & R. Ballance (Ed.), 1996. Water quality Monitoring: A practical guide to the design and implementation of freshwater quality studies and monitoring programmes. E&FN DPON.

Batchelor, C., Lovell, C, Chilton J. & Mharapara, I., 1996. *Development of collector well gardens.* 22nd WEDC Conference: Reaching the unreached, challenges for the 21st century. New Delhi, India, p172-174.

Beekman HE & Sunguro S, 2002. Groundwater recharge estimation – sustainability and reliability of three types of rain gauges for monitoring chloride deposition. Groundwater Division, Western Cape Conference: Tales of a hidden treasure; Somerset west, 16 Sept 2002.

Beekman, H. E. & Sunguro, S., 2002. *Groundwater recharge estimation – suitability and reliability of rain gauges for monitoring chloride deposition.* Groundwater Division, Western Cape conference: Tales of a hidden treasure, Somerset West, S. Africa.

Beekman, H. E., Gieske, A and Selaolo, E. T. 1996. GRES: *Groundwater recharge studies in Botswana* (1987-1996), Botswana Journal of Earth Science, Volume 3, 1-17.

Best, M. G., 1979. *Igneous and metamorphic petrology.* Freeman and Company.

Böhlke J-K, 2002. Groundwater recharge and agricultural contamination. Hydrogeol J. 10:153-179.

Boonstra J & Bhutta MN, 1996. Groundwater recharge in irrigated agriculture: theory and practice of inverse modeling. J Hydrol. 174:357-374.

Botha, L.   J and Bredenkamp, D. B. 1993. Lichtenburg: A case study incorporating cumulative rainfall departures (CRD). A case study incorporated in the manual on: Manual on quantitative estimation of groundwater recharge and aquifer storativity, Water Research Commission, Pretoria, 1-25.

Botha, L. J. 1994. Estimation of the ZeerustlRietpoort groundwater potential. A case study incorporated in the manual on: Manual on quantitative estimation of groundwater recharge and aquifer storativity, Water Research Commission, Pretoria, 1-25.

Brassington, R., 1988. *Field Hydrogeology.* 2nd Ed. John Wiley & Sons.

Bredenkamp, D. B., 1978. Quantitative estimation of groundwater recharge with special reference to the use of natural radioactive isotopes and hydrological simulation. Technical Report no 77, Department of Water Affairs, South Africa.

Bredenkamp, D.B., Botha, L. T., Van Tonder, G. J and Van Rensburg, H. J. 1995. *Manual on quantitative estimation of groundwater recharge and aquifer storativity,* Water Research Commission, Pretoria, pp. 363.

Bresler E, 1981. *Transport of salts in soils and subsoils.* Agric. Water Manage., 4:35-61.

Butterworth, J.A., F., Mugabe, L.D. Simmons and M. G. Hodnett, 1999a. Hydrological processes and water resources management in a dryland environment II: Surface redistribution of rainfall within fields. Hydrol. Earth System Sci., 3, 333-343.

Butterworth J. A., D. M. J. Macdonald, J. Bromley, L. P. Simmonds, C. J. Lovell and F. Mugabe, 1999b. Hydrological processes and water resources management in a dryland environment III: Groundwater recharge and recession in a shallow aquifer. Hydrology and Earth System Sciences, 3(3), 345-352.

Calf, G. F., 1978. An investigation of recharge to the Namoi Valley aquifers using environmental isotopes. Aust. J. Soil Res. 16, 197-207.

Caro, R & Eagleson, P. S. 1981. *Estimating aquifer recharge due to rainfall.* J. Hydrol. 53, 185-211.

Chenje M., Johnson P., 1996. *Water in Southern Africa.* SADC/IUCN/SARDC Report.

Chilton, P. J. & Foster, S. S. D., 1995. Hydrogeological characterisation and water supply potential of basement aquifers in tropical Africa. Hydrogeology Journal 3 (1), p36-49.

Chilton, P. J. and A.K. Smith-Carrington, A.K., 1984. Characterisation of the weathered basement aquifer in Malawi in relation to rural water-supplies: IAHS Publication no. 144, p. 57-72.

Committee for hydrological research. T.N.O., 1966. *Water balance studies.*

Cook PG, Hatton TJ, Pidsley D, Herczeg AL, Held A, O'Grady A and Eamus D, 1998. Water balance of a tropical woodland ecosystem, Northern Australia: a combination of micro-meteorological, soil physical and groundwater chemical approaches. J Hydrol. 210:161-177.

Cook, PG. & Walker GR., 1995. *An evaluation of the use of 3H and 36 Cl to estimate groundwater recharge in arid and semi-arid environments.* In: Proc IAEA Int Symp on Isotopes in Water resources Management, Vienna, pp 20-24.

Custodio, C., 2002. *Aquifer overexploitation: what does it mean?* Hydrogeol. J. 10:254-277.

Dahmen ER and Hall MJ, 1990. Screening of Hydrological data: Tests for Stationarity and Relative Consistency. ILRI Publication No 49.

Davisson ML, 2000. Discussion regarding sources and ages of groundwater in Southeastern California. US Dept. of Energy.

De Groen, M. M., 2002. Modeling Interception and Transpiration at Monthly Time Steps - Introducing Daily Variability through Markov Chains. PhD Thesis, IHE Delft/Technical University of Delft.

De Laat & Savenije, 1994. Principles of Hydrology. IHE Lecture Note.

De Ridder, 1972. Groundwater Resources, Final report. FAO, Rome, 219 pp.

De Silva RP, 1999. Estimating groundwater recharge in the dry zone of Sri Lanka using weekly, 10-daily or monthly evapotranspiration data. Journal of Environmental Hydrology 7:paper 4.

De Vries, J J & Simmers, I., 2002. *Groundwater recharge: an overview of processes and challenges.* Hydrogeol. Jo. (2002) 10:5-17.

De Vries, JJ, Selaolo ET, Beekman HE., 2000. Groundwater recharge in the Kalahari, with reference to paleo-hydrologic conditions. J. Hydrol. 238:110-123.

DRSS (Department of Research and Specialist Services), 1992. *Communal land physical inventory – Mhondoro and Ngezi*. Report No A551.

Doorenbos J & Pruitt WO, 1992. *Crop water requirements*. FAO Irrigation and Drainage paper no 24.

Edmunds WM & Gaye CB, 1994. Estimating the variability of groundwater recharge in the Sahel using chloride. J Hydrol. 156:47-59.

Edmunds WM, Fellman E, Goni IB & Prudhomme C, 2002. *Spatial and temporal distribution of recharge in northern Nigeria*. Hydrogeolo. J. 10:205-215.

Edmunds, W.M., Darling, W.G. and Kinniburgh, D.G., 1988. *Solute profile techniques for recharge estimation in semi-arid terrain*, in Simmers, I. Ed., Estimation of natural recharge: NATO Series C, v 222, pp 139-158.

Everson CS, 2001. The water balance of a first order catchment in the montane grasslands of S. Africa. J Hydrol. 241:110-123.

Farquharson, F. A. & Bullock, A., 1992. *The hydrology of basement complexes of Africa with particular reference southern Africa*: Geological Society special publication no 66. p59-76.

Fetter, C. W., 1994. *Applied Hydrogeology*. Prentice Hill.

Forster, S.S.D., 1984. African groundwater development – the challenges for hydrological science: IAHS Publication no. 144, p3-12.

Foster, S. S. D., Bath, A. H., Farr, J. L.& Lewis, W. J., 1982. *The likelihood of groundwater recharge in the Botswana Kalahari*. J. Hydrol. 55, 113-136.

Gbedzi VD, 1996. Selection of sub-catchments and design of monitoring network to analyze the influence of land use change on the hydrology of the Mupfure Basin. MSc Thesis HH270, IHE Delft.

Gieske ASM & De Vries JJ, 1990. Conceptual and computational aspects of the mixing cell method to determine groundwater recharge components. J Hydrol 121:277-292.

Gieske ASM, 1992. Dynamics of groundwater recharge: a case study in semi-arid eastern Botswana. PhD Thesis, Vrije Universiteit, Amsterdam, 289 pp.

Gleick, P.H. (Ed.), 1993. Water in a crisis – a guide to the world's fresh water resources. Oxford University Press.

Gonfiantini, R., Conrad, G., Fontes, J. C. H., Suazay, G. & Payne, B. R., 1974. *Isotopic study of the continental intercalaire aquifer and its relationship with other aquifers of the northern Sahara*. In: Isotope hydrology 1974, Symp. On isotope technioques in groundwater hydrology, Vienna, Vol. I, 227-241.

Grismer ME, Bachman S and Powers T, 2000. A comparison of groundwater recharge estimation methods in a semi-arid, coastal avocado and citrus orchard (Ventura County, California). Hydrol. Process. 14:2527-2543.

Lynch JA, Horner KS and Grimm JW, 2001. *Atmospheric deposition: temporal and spatial distribution in Pennsylvania*. ERRI, Pennsylvania State University.

Groundwater: Past achievements and future challenges. Proceedings of the XXX IAH Congress in groundwater, A. A. Balkema, Rotterdam, 313-318.

Gustafson, G. & Krasny, J., 1994. *Crystalline rock aquifers: their occurrence, use and importance*. Applied Hydrogeol. (1994) 2: 64-75.

Haase P, Pugnaire FI, Fernández EM, Puigdefábregas J, Clark SC and Incoll LD, 1996. An investigation of rooting depth of the semi-arid shrub Retama sphaerocarpa (L.) Boiss. By labeling of groundwater by a chemical tracer. J Hydrol 177:23-31

Hamill L. & Bell F.G., 1986. *Groundwater Development*. Butterworth, London.

Hendrickx JMH & Walker GR, 1997. *Recharge from precipitation*. In Simmers, I. (Ed), 1997. *Recharge of phreatic aquifers in (semi-) arid areas*. International Association of Hydrogeologists (IAH) 19. A.A. Belkema, Rotterdam. 277 pp.

Hillel, D & Tadmor, N. 1962. Water Regime and vegetation in the central Negev highlands ofIsrael. Ecology 43, 33-41.

Hirji, R., Johnson, P., Maro P., Matiza-Chiuta, T. (Eds), 2002. Defining and mainstreaming environmental sustainability in water resources management in southern Africa. SADC Technical Report.

Houston, J.F.T., 1992. *Rural water supplies: comparative case studies from Nigeria and Zimbabwe*: Geological Society Special Publication no 66: 243-257.

Howard KWF & Lloyd JW, 1978. The sensitivity of parameters in the penman evaporation equations and direct recharge balance. J Hydrol. 41:329-344.

Hudak, P. F., 2000. *Principles of Hydrogeology*. 2 Ed. Lewis Publishers. 200pp.

Jarawaza, M. 1999. Estimation of groundwater recharge - Grasslands research catchment, Marondera, Zimbabwe. MSc. Thesis, University of Zimbabwe, Harare, pp. 91.

Johnston, C. D., 1987. Preferred water flow and localised recharge in a variable regolith. J. Hydrol. 94, 129 -142.

Kennett-Smith A, Cook PG, and Walker GR, 1994. Factors affecting groundwater recharge following clearing in the south-western Murray Basin. J Hydrol 154:85-105.

Lamond RE & Leikam DF, 2002. Chloride in Kensas: Plant, soil and fertilizer considerations. Kensas State University.

Larsen F, Owen R, Dahlin T and Mangeya P, 2001. A preliminary analysis of the groundwater recharge to the Karoo formations, mid-Zambezi basin, Zimbabwe. 2nd WARFSA Symposium: IWRM: Theory, Practice Cases; Cape Town, 30-31 Oct. 2001.

Lerner, D.A., Issar, A.S.and Simmers, I. (Eds.), 1990. *Groundwater Recharge. A guide to understanding and estimating natural recharge.* International Contributions to Hydrogeology, Verlag Heinz Heise, Vol. 8: 245 pp.

Lewis, F. M. & Walker, G. R., 2002. Assessing the potential for significant and episodic recharge in southewestern Australia using rainfall data. Hydrogeol. J. 10:229237.

Llamas, R., Back, W. & Margat, J., 1992. Groundwater use: equilibrium between social benfits and potential environmental costs. Applied Hydrogeology 2/1992.

Lloyd, J. W. 1986. *A review of aridity and groundwater.* Hydrological Processed, 63-78.

Lloyd, J. W., 1994. Groundwater-management problems in the developing world. Applied Hydrogeology 4/1994.

Lloyd, J. W., Pike, J. G., Eccleston, B. L. & Chidley, T. R. E., 1987. *The hydrogeology of complex lens conditions in Quatar.* J. Hydrol.

Lloyd, J.W., 1999. Water resources of hard rock aquifers in arid and semi-arid zones. UNESCO.

Mabenge B & Nyamuranga N, 2001. Chloride deposition monitoring at Grasslands Experimental Station – Marondera. ZINWA internal report, 33 pp.

Makarau A, 1995. *Intra-seasonal Oscillatory Modes of the southern Africa summer circulation.* PhD dissertation. Uni. Of Cape Town, S. Africa. 125 pp.

Makuni, C.W. & Simango, R.Z., 1987. *Explanation of the Geological Map of the area around Norton.* Zimbabwe Geological Survey, **Short report No. 52**.

Martinelli E. & Hubert G.L., 1985.National Master Plan for Rural Water Supply and Sanitation, Vol. 2.2: Hydrogeology. Zimbabwe Government.

Martinelli, E.,1999. Guidelines for Boreholes, Groundwater use and Groundwater Monitoring. RRWD, p1-10.

Mazor E, Verhagen BTh, Sellschop JPE, Jones MT, Robins NE, Hutton LG and Jennings CMH, 1977. *Northern Kalahari groundwaters: hydrologic, isotopic and chemical studies at Orapa, Botswana.* J Hydrol. 34:203-234.

McCartney, M. P. 1998. *The hydrology of a headwater catchment containing a dambo.* PhD. Thesis, The University of Reading, pp. 266.

McCord JT, Gotway CA and Conrad SH, 1997. Impact of geologic heterogeneity on recharge estimation using environmental tracers: Numerical modelling investigation. Wat. Resour. Res. 33(6):1229-1240.

McFarlane, M.J., 1992. Groundwater movement and water chemistry associated with weathering profiles of the African surface in parts of Malawi: Geological society special publication no 66, p.101-130.

MEWRD (Ministry of Energy and Water Resources development), 1985. *Hydrology of Zimbabwe – Volume 2.1.* pp50.

MLWR (Ministry of Lands and Water Resources), 1984. An assessment of the water resources of Zimbabwe and guidelines for development planning. pp23. Zimbabwe Government.

Mjanja, R. 2000. A hydrochemical investigation of the groundwater resource around Chiweshe (upper Mazowe sub-catchment area). BSc Honours. Thesis, University of Zimbabwe, Harare, pp. 59.

Msipa C.G., 1992. The state of the environment report, Zimbabwe. Zimbabwe Government.

Mudzingwa, B., 1998. *Hard rock hydrogeology of the Nyatsime Catchment: Zimbabwe.* MSc Thesis, ITC Enschede, The Netherlands.

Mumeca, A. 1986. Effect of deforestation and subsistence agriculture on the runoff on the Kafue headwaters, Zambia. Hydrol. Sci. Jo., 31, (4), 543-554.

Muralidharan, D., Athavale, R. N. & Murti, C. S. 1988. Comparison of recharge estimates from injected tritium technique and regional hydrological modeling in the case of the of a granitic basin in semi-arid India. In: I. Simmers (ed), 195-220.

Murungweni, Z. N., 2001. *Water reform in Zimbabwe: Water sector in a historical perspective.* MSc WREM lecture paper, University of Zimbabwe.

Musariri, M., 1998. Preliminary analysis of the Mupfure experimental catchment. IHE MSc Thesis, HH343. 72 pp.

Najjar RG, 1999. The water balance of the Susquehanna River Basin and its response to climate change. J Hydrol. 219:7-19.

Nhidza, E., 1999. Integrating Forestry Land Use In Water Resources Management: Odzani River Catchment Case. MSc Thesis D.E.W. 089, IHE Delft.

Nkotagu H, 1996. Application of environmental isotopes to groundwater recharge studies in a semi-arid fractured crystalline basement area of Dodoma, Tanzania. J African Earth Sciences 22(4): 443-457.

Nonner, J. C., 1996. *Groundwater Exploration.* IHE Lecture Note, IHE Delft.

Nonner, J.C., 1995. *Principles of groundwater exploration – part 1.* IHE Delft Lecture Note HH022/95/1, 198pp.

Olayinka AI & Sogbetum AO, 2002. Laboratory measurement of the electrical resistivity of some Nigerian crystalline basement complex rocks. Afr. Jo. Sci. and Tech. Science na d Engineering Series 3(1)93-97.

Phillips, F. M., Trotman, K. N., Bentley, H. W., Davis, S. N., & Elmore, D. 1984. Chloride-36 from atmospheric nuclear weapon testing as a hydrologic tracer in the zone of aeration in arid climates. In: Recent investigations on the zone of aeration, Proc. Int. Symp., Munich, West Germany, 47 - 56.

Pitman WV. A mathematical model for generating monthly river flows from meteorological data in Southern Africa. 2/73. Univ. of Witwatersrand, Dept. of Civil Engineering, Hydrological Research Unit, S. Africa.

Pitts, 1984. Geology for Engineers

Press F. & R. Siever, 199-. *Understanding Earth.* W.H. Freeman and Company.

Price, M., 1996. *Introducing Groundwater.* Chapmen & Hall.

Prommer H, Barry DA, and Davis GB. 2000. Numerical modeling for design and evaluation of groundwater remediation schemes. Ecological Modeling 128:181-195.

Roberts G & Harding RJ, 1996. The use of simple process-based models in the estimate of water balances for mixed land use catchments in East Africa. J Hydrol. 180:251-266.

Rushton, K. R., 1988. Numerical and conceptual models for recharge estimation in arid and semi-arid zones. In: Simmers, I. (Ed.).

Rutter AJ, Kershaw KA, Robins PC and Morton AJ, 1971. A predictive model of rainfall interception in forests. I. Derivations oof the model from observations in a plantation of Corsican pine. Agric. Meteo. 9:267-384.

Salama RB, Tapley I, Ishii T, Hawkes G, 1994. Identification of areas of recharge and discharge using Landsat-TM satellite imagery and aerial photography mapping techniques. J Hydrol 162: 119-141.

Sami K & Hughes DA, 1996. A comparison of recharge estimates to a fractured sedimentary aquifer in South Africa from a chloride mass balance and an integrated surface-subsurface model. J Hydrol 124: 229-241.

Sanford, W., 2002. *Recharge and groundwater models: an overview.* Hydrogeol. Jo. 10:110-120.

Sandström, K., 1995. Forests and water – friends or foes? Hydrological implications of deforestation and land degradation in semi-arid Tanzania. PhD Thesis. Linkoping University.

Sangwe, K. M., 2001. Recharge map of Zimbabwe and validation of rain gauges for country wide total atmospheric chloride deposition studies. MSc WREM Thesis, Dept. of Civil Engineering, Uni. Of Zimbabwe.

Savenije H.H.G., 1995. New definitions for moisture recycling and the relationship with land-use changes in the Sahel. J Hydrol. 167:57-78.

Savenije H.H.G., 1997. Determination of evaporation from a catchment waster balance at a monthly time scale. Hydrol. And Earth System Sciences 1:93-100.

Savenije, H.H.G., 1998. How do we feed a growing world population in a situation of water scarcity. In SIWI Report 3, Water: the key to socio-economic development and quality of life. Stockholm. pp49-58.

Savenije H.H.G., 2004. The importance of interception and why we should delete the term evapotranspiration from our vocabulary. . Hydrological Processes, 18(8):1507-1511.

Savenije H.H.G., 2005. Interception. Article SW-460, The Encyclopedia of Water, Wiley.

Sekhar, M., Kumar, M.S.M. and Sridharan, K., 1994. *A leaky aquifer model for hard rock aquifers.* Applied Hydrogeol. 3/94, p32-39.

Selaolo, E. T. 1998. Tracer studies and groundwater recharge assessment in the eastern fringe of the Botswana Kalahari - The Letlhakeng-Botlhapatlou area. Ph.D. Thesis, Free University Amsterdam, pp. 229.

Selaolo, E. T., Hilton, D. R and Beekman, H. E. 2000. Groundwater recharge deduced from He isotopes in Botswana Kalahari. In: Sililo et al. (editors),

SGO (Surveyor Generals's Office), 1984. 1:25,000 Aerial Photos.

SGO (Surveyor Generals's Office), 1990. 1: 50,000 topographical map.

Sharma ML & Hughes MW, 1985. Groundwater recharge estimation using chloride, deuterium and oxygen-18 profiles in the deep coastal sands of western Australia. J Hydrol. 81:93-109.

Shaw EM, 1994. *Hydrology in Practice. 3rd Edition.* Van Nostrand Reinhold (International), London.

Silberstein RP, Sivapalan M and Wyllie A, 1999. On the validation of a coupled water and energy balance model at small catchment scales. J Hydrol. 220:149-168.

Simmers, I. (Ed), 1997. *Recharge of phreatic aquifers in (semi-) arid areas.* International Association of Hydrogeologists (IAH) 19. A.A. Belkema, Rotterdam. 277 pp.

Simonic, M., Adams, A., Carlsson, L and Marobela, C. 2000. Estimation of groundwater recharge to the Kanye dolomite aquifer in Botswana. In: Sililo et al. (editors), Groundwater: Past achievements and future challenges. Proceedings of the XXX IAH Congress in groundwater, A. A. Balkema, Rotterdam, 325-330.

Smit GN & Rethman NFG, 2000. The influence of tree thinning on the soil water in a semi-arid savanna of southern Africa. Jo of Arid Environments 44:41-59.

Smith-Carrington, A. K and Chilton , P. J. 1983. *Groundwater resources of Malawi*, Overseas Development Administration: Institute of Geological Sciences, United Kingdom, pp 172.

Sokolov, A. A. & Chapman, T. G., 1974. Methods for water balance computations. An international guide for research and practice. UNESCO Press, Paris.

Sophocleous MA, 1992. Groundwater recharge estimation and regionalisation: the Great Bend Prairie of central Kensas and its recharge statistics. J Hydrol. 137:113-140.

Sophocleous MA, 1997. Managing water resources systems: Why safe yield is not sustsinable. Ground Water 35 (4), 561.

Sophocleous MA, 2000. From Safe Yield to sustainable development of water resources – the Kansas experience. J Hydrol 235: 27-43.

Sophocleous MA & Perkins SP, 2000. Methodology and application of combined watershed and groundwater models in Kensas. J Hydrol 236: 185-201.

Stagman, J. G., 1978. *An outline of the geology of Rhodesta.* Rhodesia Geological Surrvey Bulletin 80:1-126.

Sukhija, B.S. & Rao, A. A., 1983. Environmental tritium and radiocarbon studies in the Vedavati River Basin, Karnataka and Andra Pradesh, India. J. Hydrol. 60, 185196.

Taniguchi M & Sharma ML, 1993. Determination of groundwater recharge using the change in soil temperature. J Hydrol. 148: 219-229.

Taylor, R. & Barrett, M., 1999. *Urban groundwater development in sub Saharan Africa.* 25th WEDC Conference Proc., Integrated development for water supply and sanitation. Addis Ababa, Ethiopia, 1999, p203-207.

Taylor, R. & Howard, K., 2000. A tectono-geomorphic model of the hydrogeology of deeply weathered crystalline rock: Evidence from Uganda. Hydrogeology Journal 8(3); 279-294.

Thiery, D. 1988. Analysis of long duration piezometric records from Burkina Faso to determine aquifer recharge. In: I. Simmers (ed), 477-489.

Todd, D. K. 1980. Groundwater Hydrology.

Torrance JD, 1981. *Climate handbook of Zimbabwe.* 551.582(689.1), Zimbabwe Dept. Of Meterological Services, 220 pp.

Tucker ME, 1991. Sedimentary Petrology: An introduction to the origin of sedimentary rocks. 2nd Ed. Blakewell Scientific Publications.

Van der Lee J & Gehrels JC, 1990. *Modeling aquifer recharge. Introduction to the lumped parameter model EARTH.* Inst. Earth Sci., Vrije Univ., Amsterdam/Geol. Survey, Botswana. 29 pp.

Van der Lee J & Gehrels JC, 1997. *Modeling of groundwater recharge for a fractured dolomite aquifer under semi-arid conditions.* In Simmers I (Ed) Recharge of phreatic aquifers in (semi) arid areas. IAH Int Contrib Hydrogeology 19. AA Balkema Rotterdam, pp 129-144.

Van der Moot, N. L., 2001. Exercises and Procedures for the Lecture: Interpretation of Geo-electrical Measurements. Including Case Study Breevenen. IHE lecturenote HH454/01/1

Van Genuchten MTh, Leij FJ and Lund LJ (eds), 1992. *Indirect methods for determining the hydraulic properties of unsaturated soils.* In Proc 1989 Int Worksh, California, 718 pp.

167

Van Tonder, G., 1999. Groundwater management under the new National Water in South Africa. Hydrogeol. J. 7: 421-422.

Verhagen B.Th., 1984. Environmental Isotope study of a groundwater study in the Kalahari of Gordonia. Isotope Hydrology 1983, Proc Symp. IAEA, 415-433.

Wagner, W., Hobler, M., Kohler, G and Giesel, W. 1987. Lomagundi aquifer study: Groundwater use and groundwater potential in the Chinhoyi - Umboe Mhangurafarming areas, Hanover.

Walker GR, Jolly ID and Cook PG, 1991. A new chloride leaching approach to the estimation of diffuse recharge following a change in land use. J Hydrol 128: 49-67.

Ward RC & Robinson M, 1990. *Principles of Hydrology*. 3rd Edition. McGraw-Hill, London.

William WGY, 1986. Review of parameter identification procedures in groundwater hydrology: the inverse problem. Wat. Resourc. Res. 22(2) 95-108.

Wilson, W., 1969. *Engineering Hydrology*. McGraw Hill.

WCED (World Commission on Environment and Development), 1987. *Our Common Future*. Oxford University Press, New York. NY.383pp.

Worst, B.G.,1962. *The Geology of the Mwenesi Range and the adjoining country*. Southern Rhodesia Geological Survey Bulletin No. 54.

Wright, E. P., 1992. *The hydrogeology of crystalline basement aquifers in Africa*: Geological society special publication no 66, p1-28.

WRRC (Water Resources Research Centre), 1980. *Regional recharge research for southeast alluvial basins. Final report on USGS contract 14-08-0001-18257*. Department of Hydrology and Water Resources, University of Arizona, Tuscon. 417 pages.

Wurzel, P., 1983. *Tritium as a groundwater tracer in Zimbabwe*. Methods and instrumentation for the investigation of groundwater systems, 289-300.

Xiong L & Guo S, 1998. A two parameter monthly model and its application. J Hydrol. 216:111-123.

APPENDIX: NOMENCLATURE & ABBREVIATIONS

List of symbols

Symbol	Description	Dimension
A	surface area	[L2]
C	capillary rise	[L]
C_m	maximum capillary rise	[L]
c_g	groundwater chloride concentration	$[ML^{-3}]$
c_p	chloride concentration in rainfall	$[ML^{-3}]$
d	dry deposition as fraction of wet deposition	[-]
dh	change in water level	[L]
dS	change in catchment storage	[L]
D_s	dry deposition, chloride	$[ML^{-2}T^{-1}]$
D_r	root depth	[L]
D_{wt}	depth to the water table	[L]
E	total evaporation	$[LT^{-1}]$
E_s	evaporation from the soil surface	$[LT^{-1}]$
E_o	open water evaporation	$[LT^{-1}]$
F	infiltration	$[LT^{-1}]$
h	water level	[L]
h_p	height of water column from pressure sensor	[L]
h_p	daily per capita water consumption	[L]
I	interception	$[LT^{-1}]$
I_p	potential daily interception	$[LT^{-1}]$
I_r	rainfall intensity	$[LT^{-1}]$
I_{rmax}	maximum rainfall intensity	$[LT^{-1}]$
I_s	interception from the ground surface	$[LT^{-1}]$
I_v	vegetation canopy interception	$[LT^{-1}]$
K_{gb}	time scale of slow groundwater	[T]
K_{gf}	time scale of fast groundwater	[T]
k	hydraulic conductivity	$[LT^{-1}]$
k_h	horizontal hydraulic conductivity	$[LT^{-1}]$
k_v	vertical hydraulic conductivity	$[LT^{-1}]$
L	length dimension	[L]
M	mass dimension	[M]
N	quantity	[-]
n	counter limit	[-]
P	rainfall	$[LT^{-1}]$
P_a	atmospheric pressure	[L]
P_{net}	net rainfall	$[LT^{-1}]$
P_w	water pressure	[L]
Q_o	total catchment stream discharge	[L/T]
Q_s	surface runoff	[L/T]
Q_{gb}	slow groundwater discharge (baseflow)	[L/T]
Q_{gf}	fast groundwater discharge (fast flow)	[L/T]
R	recharge	[L/T]
S	total catchment storage	[L]
S_{gm}	maximum catchment groundwater storage	[L]
S_s	surface storage	[L]
S_o	open water storage	[L]

169

S_u	unsaturated zone storage	[L]
S_g	saturated zone storage	[L]
S_r	specific retention	[-]
S_y	specific yield	[-]
T	time dimension	[T]
T	time period	[T]
t	time step	[T]
t_o	time at start of a process	[T]
ϕ	porosity	[-]
μ	diffusion coefficient	

Units

masl	metres above sea level
mbgl	metres below ground level
mm/a	millimeters per annum
mm/d	millimeters per day
mm/hr	millimeters per hour
mm/month	millimeters per month
Mm^2	square kilometers (million square metres)
Ohm.m	ohm metre

Subscripts

fc	field capacity
g	saturated zone
h	horizontal direction
i	counter
max	maximum
min	minimum
eff	effective
net	net
o	catchment total
ow	open water
p	potential
pd	rainfall, dry
pw	rainfall, wet
r	root
s	surface zone
t	time
u	unsaturated zone
v	vertical direction
veg	vegetation
wp	wilting point

Abbreviations

ARF   Areal Reduction Factor
BFI   Baseflow index
CRD   cumulative rainfall departure
CMB   chloride mass balance
CBA   crystalline basement aquifer
CEC   Cation exchange capacity
CSAC   cumulative storage accumulation
CV   coefficient of variation
EARTH
ENB   Electrical neutrality balance
EPS   Electrical Profiling Survey
FMB   fluid mass balance
GPS   Global Positioning System
IHS   instantaneous hydrograph separation
IOC   Indian Ocean Cyclones
ITCZ   Inter-Tropic Convergence Zone
IUGC   International Union of Geological Classifications
SADC   Southern African Development Community
SMB   soil moisture balance
SVF   saturated volume flux
USDI   United States Department of the Interior
VES   Vertical Electrical Sounding
WB   water balance
WTF   water table fluctuations

Terminology

Recharge efficiency   the recharge contribution from total rainfall, R/P.
Short-term   planning horizon below three years
Long-term   planning horizon of ten or more years
Study period   duration of this study
Long record   data collected over a period of ten years or more
Short record   data collected over a period of ten years or less
Recharge period   the part of the rain season when recharge is highest
Blue water   all water that can be accessed through engineering interventions
White water   evaporation from interception and wet surfaces
Green water   part of rainfall used for biomass production

T - #0010 - 071024 - C182 - 280/208/20 [22] - CB - 9780415416924 - Gloss Lamination